DISASTER AND EMERGENCY PLANNING FOR PREPAREDNESS, RESPONSE, AND RECOVERY: PROMOTING RESILIENT INFRASTRUCTURE AND COMMUNITY

WEF Special Publication

Second Edition

Edited by Lucy Lunevich

2021

Water Environment Federation
601 Wythe Street
Alexandria, VA 22314-1994 USA
www.wef.org

ISBN: 978-1-57278-413-0

About WEF

The Water Environment Federation (WEF) is a not-for-profit technical and educational organization of 30,000 individual members and 75 affiliated Member Associations representing water quality professionals around the world. Since 1928, WEF and its members have protected public health and the environment. As a global water sector leader, our mission is to connect water professionals; enrich the expertise of water professionals; increase the awareness of the impact and value of water; and provide a platform for water sector innovation. To learn more, visit www.wef.org.

Second edition edited by Lucy Lunevich, PhD

With contributions by Brendan Moon

Review provided by:

Steve Mustard, PE, CEng, CAP
Vishakha Kaushik, PE
Andrew Ohrt, PE
Kenneth C. Morgan, PE
Bryan A. Mantz, CMC, CGFM
Joe Paterniti

Text originally prepared by the **Emergency Planning, Response, and Recovery Task Force** of the **Water Environment Federation**

James J. Newton, PE, BCEE, *Chair*

Archis Ambulkar, MS
John Bannen
Joseph R. Becker, PE
Jonathan Boitano, PE
Kari Fitzmorris Brisolara, ScD, MSPH
Al Callier, PE
Chein-Chi Chang, PhD, PE
Shellie Chard-McClary
Laurie Chase, PE
Mark Christison
Michael Davis
R. Ryan Dupont, PhD
Jim Faccone
Richard E. Finger
Gary D. Fournier
Steve Frank
Wes Frye
Robin J. Gruenfeld
Prof. Francis J. Hopcroft, PE
Nicole L. Ivers, CIH
Hari Kapalavai, PE, BCEE
Hiroyuki Kato
Kanwer Khan
Mayumi Koseki
John Lawson, RN, MN
Lucy Lunevich, PhD
Garry Macdonald
Manabu Matsuhashi
Isha Matta
Herbert Miller
Jack Moyer
Tim Page-Bottorff, CSP
Tom Pedersen, ENV SP
Jonathan P. Reeves
Smita Sivakumar
Marc Walch
John H. Webber

Under the Direction of the **Safety, Occupational Health, and Security Subcommittee** of the **Technical Practice Committee**

Special Publications of the Water Environment Federation

The WEF Technical Practice Committee (formerly the Committee on Sewage and Industrial Wastes Practice of the Federation of Sewage and Industrial Wastes Associations) was created by the Federation Board of Control on October 11, 1941. The primary function of the Committee is to originate and produce, through appropriate subcommittees, special publications dealing with technical aspects of the broad interests of the Federation. These publications are intended to provide background information through a review of technical practices and detailed procedures that research and experience have shown to be functional and practical.

Water Environment Federation Technical Practice Committee Control Group

Contents

List of Figures xix
List of Tables xxi
Preface xxiii

Chapter 1 Introduction 1
1.0 OVERVIEW 2
1.1 Scope and Purpose 2
1.2 Definitions 4
1.2.1 Hazards 4
1.2.2 Disasters 4
1.2.3 Vulnerability 4
1.2.4 Resilience 5
1.3 Organization 6
1.3.1 Planning 6
1.3.2 Response 6
1.3.3 Recovery 6
1.3.4 Case Studies 7
2.0 EMERGENCY SITUATIONS AND POTENTIAL EFFECTS ON WASTEWATER UTILITIES 7
2.1 Infrastructure Failure 7
2.1.1 Industrial Accidents 7
2.1.1.1 Conveyance System Failures 7
2.1.1.2 Fires 8
2.1.1.3 Hazardous Chemicals 8
2.1.1.4 Radiation Release 8
2.1.2 Electrical Outages 9
2.1.2.1 Community 9
2.1.2.2 Facility 9

2.1.3 Communications 9

2.1.3.1 Continuity Planning 9

2.1.3.2 Technological Failures 10

2.1.4 Supply Chain 10

2.1.5 Transportation Accidents 11

2.2 Natural Disasters 11

2.2.1 Biological 11

2.2.2 Geophysical 12

2.2.2.1 Earthquakes 12

2.2.2.2 Volcanic Eruptions 12

2.2.2.3 Landslides 13

2.2.2.4 Avalanches 13

2.2.3 Hydrological 13

2.2.3.1 Floods 14

2.2.3.2 Tsunamis and Tidal Waves 14

2.2.3.3 Mudslides 14

2.2.4 Meteorological 14

2.2.4.1 Hurricanes 15

2.2.4.2 Tornadoes 15

2.2.4.3 Lightning Strikes 15

2.2.5 Climatological 16

2.2.5.1 Dust Storms 17

2.2.5.2 Extreme Temperatures 17

2.2.5.3 Wildfires 17

2.2.5.4 Drought 17

2.3 Human-Induced Incidents 18

2.3.1 Personnel Issues 18

2.3.2 Civil Unrest 19

2.3.3 Criminal Activities 19

2.3.4 Terrorism 20

2.4 Multihazard Disasters 22

3.0 DISASTER PLANNING AND RESPONSE 22

3.1 Preparedness 23

3.2 Response 25

3.3 Recovery 27

4.0 SUMMARY 27
5.0 REFERENCES 28

Chapter 2 Emergency Planning 35
1.0 REGULATORY REQUIREMENTS 36
1.1 Emergency Planning Process 36
1.1.1 Why Plan for Emergencies? 36
1.1.2 Who Should Plan? 37
1.1.3 Resource Conservation and Recovery Act (Contingency Planning) 39
1.1.4 Emergency Plans Required by Regulations 39
1.1.5 Facility Response Plan 39
1.1.6 Spill Prevention, Containment, and Countermeasure Plans 41
1.1.7 Oil Removal Contingency Plan 42
1.1.8 Stormwater Pollution Prevention Plan 42
1.1.9 Resource Conservation and Recovery Act Contingency Plans 43
1.1.10 Cybersecurity 44
1.1.11 Risk Management Plan 44
1.1.12 Training and Community Awareness 45
1.1.13 Building Disaster-Resilience Community 45
1.2 Emergency Response Center and Operation 46
1.2.1 Introduction 46
1.2.2 Command and Control Structure 48
1.2.3 Incident Command System 49
1.2.4 Communication and Information Management 49
1.2.5 Decision Process 51
1.2.6 Organization Structure 52
1.2.7 Authority 52
1.2.8 Media Involvement 53
1.3 Risk Management Plans 54
1.3.1 Disaster Prevention and Mitigation Engineering 54
1.3.2 Communication and Consultation 55
1.3.3 Establish Context 57

1.3.4 *Risk Identification* 58
1.3.5 *Risk Analysis* 58
1.3.6 *Risk Evaluation* 59
1.3.7 *Risk Treatment* 59
1.3.8 *Monitoring and Reviewing* 61
1.3.9 *Vulnerability Assessment* 62
1.3.10 *Risk Perception* 66
1.4 Occupational Safety and Health Administration Emergency Action Plans 66
1.4.1 *Introduction* 66
1.4.2 *Procedures for Reporting* 66
1.4.3 *Procedures for Emergency Evacuation* 68
1.4.4 *Procedures for Employees Who Remain to Operate* 68
1.4.5 *Employee Alarm System* 69
1.4.6 *Staff Training* 69
1.5 Occupational Safety and Health Administration Fire Prevention Plans 70
1.5.1 *Significant Fire Hazards* 70
1.5.2 *Procedures to Control Combustible Waste Materials* 71
1.5.3 *Procedures for Regular Maintenance of Heat-Producing Equipment* 71
1.5.4 *Staff Training* 71
2.0 DEVELOPING AN EMERGENCY RESPONSE PLAN 72
2.1 Introduction 72
2.1.1 *Emergency Planning Process* 73
2.1.2 *Elements of a Plan* 74
2.1.3 *Telecommunication Destruction* 76
2.1.4 *Resources and Integration of Agencies* 76
2.1.5 *Assessment of Community Resiliency* 77
2.2 Exercising the Plan 77
2.2.1 *Internal and External Emergency Planning* 77
2.2.2 *Planning Process* 78
2.2.3 *Exercises and Training* 79
2.3 Revising the Plan 80

3.0 REFERENCES 80
4.0 SUGGESTED READINGS 81

Chapter 3 Disaster Response 87
1.0 INTRODUCTION 88
2.0 FEDERAL AND REGIONAL RESPONSE SUPPORT 91
2.1 Department of Homeland Security National Response Framework 91
2.2 National Incident Management System 94
2.3 Incident Command System 94
2.4 National Preparedness Guidelines 96
2.5 Federal Response Support—U.S. Environmental Protection Agency 96
2.6 Federal Response Support—Other Agencies 97
2.6.1 Federal Bureau of Investigation 97
2.6.2 Department of Homeland Security 98
2.6.3 Centers for Disease Control and Prevention 98
2.7 Regional Support 99
2.7.1 Water/Wastewater Agency Response Network 99
2.7.2 Risk Management Planning 100
2.7.3 Incident Command Team 101
2.8 Communication 101
2.8.1 Communication Training 102
2.8.2 Contact Lists and Key Documents 103
2.8.3 Maintaining Credibility 104
3.0 UTILITIES—RESPONSE 105
3.1 Initial Response Actions 105
3.2 Planning as Incident Unfolds 106
3.3 Initial Response 106
3.3.1 The Start of Each Planning Cycle 116
3.3.2 Preparing for and Conducting the Tactics Meeting 118
3.3.3 Preparing for the Planning Meeting 119
3.3.4 Conducting the Planning Meeting 119
3.3.5 Incident Action Plan Preparation and Approval 119

3.3.6 Operations Period Briefing 120
3.3.7 Execute Plan and Assess Progress 120
3.4 Incident-Specific Procedures 120
4.0 INCORPORATING SAFETY TO DISASTER RESPONSE 124
5.0 COMMUNICATION: PLANNING BECOMES DOING 125
5.1 Say What Can Be Said 126
5.2 Create a Holding Statement 126
5.3 Tell More of What Can Be Said 128
5.4 Prepare for Two-Way Conversations; Social Media Can Be a Two-Edged Sword 128
5.5 Express Concern for Victims 128
5.6 Respond to Questions the Media Will Ask 129
5.7 Avoid Saying "No Comment" 129
5.8 Identify Affected Parties 129
5.9 Prepare Written Statements 130
5.10 Conduct a Media Briefing 131
5.11 Take Photographs 131
5.12 Additional Tips 132
6.0 RESOURCES 133
7.0 REFERENCES 153

Chapter 4 Disaster Recovery 155
1.0 INTRODUCTION—WHAT IS DISASTER RECOVERY? 157
1.1 Reversing the Trend of Declining Community Resilience 159
2.0 PLANNING AND MANAGING SMALL-SCALE AND COMMUNITYWIDE DISASTER RECOVERY 159
2.1 Small-Scale Disaster Recovery 160
2.1.1 Utility Integrated Contingency Plan Primary Objectives 161
2.1.2 Advantages of the Utility Integrated Contingency Plan Approach 161
2.1.3 Planning Characteristics of Utility's Integrated Contingency Plan 161
2.2 Communitywide Recovery 162

2.3 The Importance of Recovery Preparation and Planning 162
2.3.1 Establishing the Recovery Planning Team 162
2.3.2 Recovery Planning Process 164
2.3.2.1 Recovery Planning and Coordination With Local Plans 165
2.3.2.1.1 Public Works Plans and Policies 165
2.3.2.1.2 Emergency Management Plans and Policies 166
2.3.2.1.3 Economic Development Plans and Policies 166
2.3.2.1.4 Finance Plans and Policies 166
2.3.2.1.5 Other Local Plans and Policies 166
2.3.2.2 Recovery Planning—Staffing Needs 167
2.3.2.3 Recovery Planning—Legal Issues and Tasks 167
2.3.2.4 Recovery Planning—Finance Issues, Priorities, and Tasks 168
2.3.2.5 Documentation Requirements and Priorities 169
2.3.3 Planning for Critical Recovery Operations 169
2.3.3.1 Damage Assessment 169
2.3.3.2 Damage Assessment Memoranda of Understanding With Surrounding Utilities and Communities 170
2.3.3.3 Debris Management Planning and Multijurisdictional Memoranda of Understanding 170
2.3.3.4 State Department of Transportation Memoranda of Understanding 170
2.3.4 The Recovery Plan 171
2.3.5 Planning for Resource Needs 171
2.3.5.1 State and Local Recovery Resources 171
2.3.5.2 Local Recovery Funding 172
2.3.5.3 Few Disasters Are Supported by Presidential Declarations 172

2.3.5.4 *Local Recovery Funding Options* 173
2.3.5.4.1 *Reserve Funds* 173
2.3.5.4.2 *Risk Abatement and Risk Mitigation* 173
2.3.5.4.3 *Risk Spreading* 173
2.3.5.4.4 *Risk Acceptance* 174
2.3.6 *Establishing Recovery Priorities: Conducting Damage Assessments* 174
2.3.7 *Why Plan for Recovery Damage Assessment?* 174
2.3.8 *Information Gathered by the Recovery Task Force and Others* 174
2.3.8.1 *Using Hazard Analysis Information to Plan for Recovery Implementation* 175
2.3.8.2 *Critical Infrastructure and Key Resources Risk Assessment* 175
2.3.8.3 *Determining Critical Infrastructure and Key Resources Recovery Priorities* 175
2.3.8.4 *Economic Effects* 176
2.3.8.5 *Community Needs* 176
2.3.8.6 *Public Health Concerns* 176
2.3.9 *Identifying Potential Recovery Projects* 176
2.3.10 *Prioritization of Recovery Projects* 177
2.3.10.1 *Selecting Recovery Projects to Implement* 177
2.3.10.2 *Implementation Criteria* 178
2.3.10.3 *Recovery Priorities: Providing Essential Services (Maintaining Services)* 178
2.3.10.4 *Recovery Priorities: Restoring Critical Infrastructure and Key Resources* 179
2.3.10.5 *Recovery Priorities: Protecting Public Safety and Health* 179
2.3.10.6 *Recovery Priorities: Protecting the Environment* 179
3.0 RECOVERY PLAN AND RECOVERY OPERATIONS PLAN ACTIVATION 180

3.1 Inspection of Damaged Infrastructure 180
3.2 Providing Support Services 181
3.3 Recovery Coordination and Communication 181
3.3.1 Communication During Recovery 182
3.3.2 National Incident Management System 183
3.3.3 Intra-Agency Communication 183
3.3.4 Communication With the Public 183
3.3.5 Recovery Implementation 184
3.4 Recovery Training 184
4.0 NATIONAL DISASTER RECOVERY POLICY 185
5.0 REFERENCES 187
6.0 SUGGESTED READINGS 190

Chapter 5 Resilience 191
1.0 RESILIENCE IN DISASTER MANAGEMENT 191
1.1 The Case for Resilience 191
1.2 Understanding Resilience in Disaster Management 192
1.3 Integrating Resilience in Prevention and Preparedness 196
1.4 Integrating Resilience Into Response and Recovery 196
2.0 RESILIENT INFRASTRUCTURE 199
2.1 Funding Resilient Infrastructure 202
2.2 Water and Flood Mitigation Infrastructure 202
2.3 Land-Use Planning 203
2.4 Design and Construction 203
3.0 A SYSTEMATIC APPROACH TO EMBEDDING RESILIENCE IN DISASTER MANAGEMENT 203
3.1 Integrating Resilience in Prevention and Preparedness 204
3.2 Integrating Resilience in Response and Recovery 205
3.2.1 Damage Assessments: DAMRsys™ 205
3.2.2 Betterment 207
3.3 Building Resilient Infrastructure 208
4.0 REFERENCES 210
6.0 SUGGESTED READINGS 213

Index 215

List of Figures

1.1 Disaster Preparedness Strategies . . . 3
1.2 Planning, Response, and Recovery Phases . . . 23
1.3 Response Activities . . . 26
2.1 Incident Command Structure . . . 50
2.2 Risk Management Planning Process . . . 55
2.3 Risk Evaluation . . . 61
2.4 Emergency Action Plan From OSHA 1910.38(a) Standard . . . 67
3.1 Incident Life Cycle . . . 89
3.2 National Response Framework Support . . . 93
3.3 U.S. EPA Approach to a Response . . . 97
3.4 Planning Cycle . . . 117
3.5 Safety Message/Plan . . . 126
4.1 Infrastructure Resilience Properties . . . 160
5.1 Incorporating Resilience Into the PPRR Model . . . 193
5.2 Interdependencies That Influence Resilience . . . 194
5.3 Resilient Queensland . . . 205
5.4 Conducting In-Field Damage Assessments . . . 206
5.5 Geospatial Overlay of Damage Assessment Data in Townsville, Queensland, Following North and Far North Queensland Monsoon Trough Event, 2019 . . . 207
5.6 Image Showing the Relocation of the Gayndah Water Intake to Above the Claude Wharton Weir . . . 209
5.7 Flooding Effects of the 2015 Flood on the Gayndah Water Supply Intake . . . 209

List of Tables

1.1 2011 Natural Hazard Statistics 3
1.2 Saffir–Simpson Hurricane Wind Scale 16
2.1 Regulatory Emergency Plan Programs 40
2.2 Risk Information Checklist 56
2.3 Risk Analysis 59
2.4 Qualitative Risk Evaluation 60
2.5 Organizational and Functional Involvement in Risk Treatment 62
2.6 Example of Generic Mitigation Measures 63
2.7 Risk Reduction Measures in Land Use and Planning 64
2.8 An Example of a Matrix for Scaling Consequences of Risks 65
3.1 Four Phases of Emergency Management 90
3.2 Risk Management Terms 100
3.3 Response Matrix 107
3.4 Typical Response Timeline 111
3.5 Planning During an Incident 116
3.6 Examples of Possible Hazards and Associated Incidents/Events 121
3.7 Incident-Specific Response Procedures 122
3.8 Resource List and Best Practices. 134
3.9 Industry Standards, Tools, and Best Practices 144
4.1 Disaster Recovery Ordinance Content 163
4.2 Disaster Recovery Task Force Participants 164
5.1 Roles and Responsibilities in Building Resilience Into Prevention and Preparedness 197

Preface

Written for water, wastewater, electric, and gas utility managers; operators; consulting engineers; emergency response planners and professionals; and public officials, this publication addresses the key elements of emergency planning, including regulatory requirements and how to recover from resulting emergencies and build on resilience. Disasters faced by water resource recovery facilities include infrastructure failure such as industrial accidents, electrical failures, communications, supply chain, and transportation accidents; natural disasters, including biological, geophysical, hydrological, meteorological, and climatological; human-induced accidents, such as personal issues, civil unrest, criminal activities, and terrorism; and multihazard disasters.

In this new edition, a significant contribution was made by Queensland Reconstruction Authority, Australia, focusing on the importance of resilience in disaster management (Chapter 5).

The second edition of this Special Publication was edited and updated by Lucy Lunevich.

Authors' and reviewers' efforts were supported by the following organizations:

International Society of Automation
KCM Consulting Services, LLC
Queensland Reconstruction Authority
West Yost Associates

1

Introduction

1.0 OVERVIEW 2
- 1.1 Scope and Purpose 2
- 1.2 Definitions 4
  - *1.2.1 Hazards* 4
  - *1.2.2 Disasters* 4
  - *1.2.3 Vulnerability* 4
  - *1.2.4 Resilience* 5
- 1.3 Organization 6
  - *1.3.1 Planning* 6
  - *1.3.2 Response* 6
  - *1.3.3 Recovery* 6
  - *1.3.4 Case Studies* 7

2.0 EMERGENCY SITUATIONS AND POTENTIAL EFFECTS ON WASTEWATER UTILITIES 7
- 2.1 Infrastructure Failure 7
  - *2.1.1 Industrial Accidents* 7
    - *2.1.1.1 Conveyance System Failures* 7
    - *2.1.1.2 Fires* 8
    - *2.1.1.3 Hazardous Chemicals* 8
    - *2.1.1.4 Radiation Release* 8
  - *2.1.2 Electrical Outages* 9
    - *2.1.2.1 Community* 9
    - *2.1.2.2 Facility* 9
  - *2.1.3 Communications* 9
    - *2.1.3.1 Continuity Planning* 9
    - *2.1.3.2 Technological Failures* 10
  - *2.1.4 Supply Chain* 10
  - *2.1.5 Transportation Accidents* 11
- 2.2 Natural Disasters 11
  - *2.2.1 Biological* 11
  - *2.2.2 Geophysical* 12
    - *2.2.2.1 Earthquakes* 12
    - *2.2.2.2 Volcanic Eruptions* 12
    - *2.2.2.3 Landslides* 13
    - *2.2.2.4 Avalanches* 13
  - *2.2.3 Hydrological* 13
    - *2.2.3.1 Floods* 14
    - *2.2.3.2 Tsunamis and Tidal Waves* 14
    - *2.2.3.3 Mudslides* 14
  - *2.2.4 Meteorological* 14
    - *2.2.4.1 Hurricanes* 15
    - *2.2.4.2 Tornadoes* 15

2.2.4.3 *Lightning Strikes* 15
2.2.5 *Climatological* 16
2.2.5.1 *Dust Storms* 17
2.2.5.2 *Extreme Temperatures* 17
2.2.5.3 *Wildfires* 17
2.2.5.4 *Drought* 17
2.3 Human-Induced Incidents 18
2.3.1 *Personnel Issues* 18
2.3.2 *Civil Unrest* 19
2.3.3 *Criminal Activities* 19
2.3.4 *Terrorism* 20
2.4 Multihazard Disasters 22
3.0 DISASTER PLANNING AND RESPONSE 22
3.1 Preparedness 23
3.2 Response 25
3.3 Recovery 27
4.0 SUMMARY 27
5.0 REFERENCES 28

1.0 OVERVIEW

1.1 Scope and Purpose

Disasters can disrupt the essential services provided by wastewater systems and utilities, including the protection of public health and the environment. Planning for emergencies equips utilities to better respond to and recover from these events and helps to mitigate the economic, environmental, and social toll of disasters. Developing a comprehensive disaster preparedness strategy must be undertaken considering the plethora of activities and entities that become involved in response and recovery efforts, as shown in Figure 1.1.

In disasters, water resource recovery facilities (WRRFs) and collection systems are potentially subject to damage. This damage is not isolated; the community being served also will be affected and advisories can significantly change treatment needs. For example, evacuations of the communities using the system will decrease the influent load. However, in the scenario of sheltering in place, systems manage both increased storm water and increased loading from households. In either situation, proper planning using an all-hazards approach will aid in decision-making during emergency plan development. An example of the all-hazards approach to emergency planning was instrumental in a quick and effective response by the City of Dublin, Ohio, to a sanitary sewer surcharge in 2010.

Table 1.1 provides a representation of disaster risk for 2011 provided by the National Weather Service (NWS) (2012) and the National Oceanic and Atmospheric Administration (NOAA) (2012b). Though this information is related only to natural hazards, these publications also address other

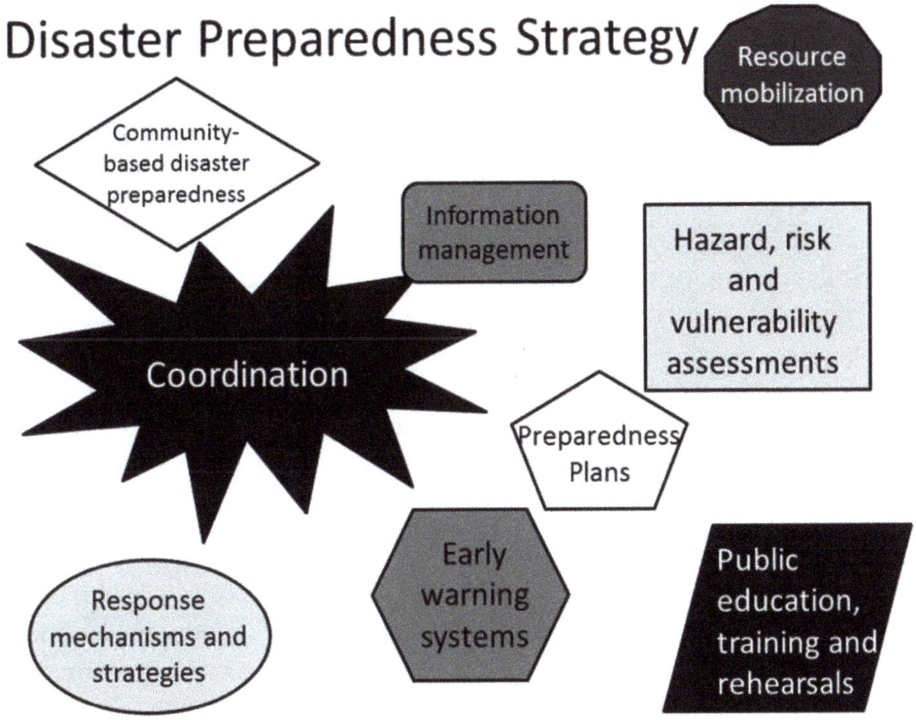

FIGURE 1.1 Disaster Preparedness Strategies

disasters, such as infrastructure failure and human-induced incidents, pertinent to utilities. This special publication provides a discussion of the elements integral to disaster preparedness strategy, including regulatory requirements, specifically focused on wastewater systems and utilities. Emergency planning, response, and recovery approaches are discussed along with case studies from around the world illustrating best practices.

TABLE 1.1 2011 Natural Hazard Statistics (NOAA, 2012b; NWS, 2012)

Weather event	Fatalities	Injuries	Total damage (million $)
Convection	635	6085	10 584.15
Extreme temperatures	235	2409	368.39
Flood	113	40	8216.9
Marine	46	40	339.13
Tropical cyclones	9	3	653.68
Winter	28	23	169.41
Other	30	230	3607.88

1.2 Definitions

1.2.1 Hazards

The International Federation of Red Cross and Red Crescent Societies (IFRC) defines a hazard as "A dangerous phenomenon, substance, human activity or condition that may cause loss of life, injury or other health impacts, property damage, loss of livelihoods and services, social and economic disruption, or environmental damage" (IFRC, 2018).

The progression of defining hazards has shifted over the past century through four phases from engineering (physical causes) to behavioral (changing behavior to minimize risk) to development (economic development's contribution to vulnerability) to complexity (sustainability and complex interaction improving long-term management of hazards) (Smith & Petley, 2008).

1.2.2 Disasters

A hazard occurrence (i.e., earthquake, flood, or hurricane/cyclone) becomes a disaster when it results in injuries, loss of life and livelihoods, displacement and homelessness, and/or destruction and damage to infrastructure and property. A disaster is an extreme disruption of the functioning of a society that causes widespread human, material, or environmental losses that exceed the ability of the affected society to cope using only its own resources. Events such as earthquakes, floods, and cyclones, by themselves, are not considered disasters. Rather, they become disasters when they adversely and seriously affect human life, livelihoods and property (IFRC, 2018). A cyclone that surges over an uninhabited island does not result in a disaster; however, it would be a disaster if it affected a populated area and caused extensive loss of lives and property. Defining disasters is difficult. Debates have raged over the proper construct of a definition, and how to incorporate the social effects with the physical, often monetary, implications (Britton, 1986; Rodríguez et al., 2007).

1.2.3 Vulnerability

Vulnerability is considered from two perspectives: structural/physical and human. Quantification of the physical or structural vulnerabilities is typically easier than determining the potential human vulnerability especially as related to the delivery of wastewater services. With the advent of cybersecurity risks, another category of vulnerability exists: computers or computer networks may have technical flaws that can be exploited intentionally or unintentionally to cause an incident that may result in serious consequences for an wastewater utility (IFRC, 2018). The physical or structural vulnerability is "the extent to which a structure is likely to be damaged or disrupted by a hazard event" (IFRC, 2018). According to the IFRC, human vulnerability is the "relative lack of capacity of a person or community to anticipate, cope with, resist and recover from the effect of a hazard. Factors that increase human vulnerability to disasters include rapid urbanization, population growth, and lack of knowledge about how effectively to resist the effects of disasters and poverty. Of all the factors, poverty is perhaps at the root of what makes most people vulnerable to the impact of most hazards" (IFRC, 2018).

1.2.4 Resilience

Resilience involves the ability to recover after stress or to adjust to dramatic changes. Dispositional resilience (hardness) represents the characteristic way in which person approaches and interprets an experience.

Resilience of infrastructure and community should be considered from economic, environmental and social perspectives. A challenge for mitigation planning and intervention to reduce vulnerability is for planners, engineers and social scientists to design and maintain infrastructure which support community physical and social resilience.

Resilience defines the active process of self-righting, learned resourcefulness and growth, the ability to function physiologically at a level greater than expected given the individual's known capabilities and previous experiences. While resilience is facilitated by certain personal traits, it can be developed over time.

The environment can foster or sustain resilience. A resilience environment is one in which people can maximize their potential for recovery and growth. The environment and infrastructure should be designed to allow the community and the individual to continue to function at an acceptable level and to mitigate any adverse consequences. A resilient environment is characterized by:

- Defining the community and infrastructure as resilient rather than vulnerable;
- Re-orienting systems to encourage learned resourcefulness through participative development programs;
- Creating vision and commitment to support others; and
- Creating a therapeutic community (building cohesion, supporting education and awareness, fostering and sustaining self-governance, control and leadership).

Infrastructure should support community needs and this is a major challenge for disaster management authority (Cronin, 2018).

1.3 Organization

1.3.1 Planning

Planning is the primary preparedness activity. The components of regulatory requirements in addition to the steps to develop an effective emergency response plan are covered in Chapter 2. The regulatory influences on emergency preparedness such as developing a contingency plan for the Resource Conservation and Recovery Act and the Occupational Safety and Health Administration's emergency action plan and fire prevention plan are discussed. The essential elements and process of developing an effective emergency response plan are addressed in Chapter 2, Section 2.0, which provides assistance in the development of the plan and exercises, training, and evaluation.

1.3.2 Response

In the response phase of emergency situations, utility managers and staff should be intimately familiar with the response plan having completed significant levels of training and exercises. For the response phase, it is essential to have access to accurate, updated information, which relies on a direct line of communication. The key components of an effective response plan are covered in the Federal Emergency Management Agency's (FEMA's) Incident Command System (ICS) and include incident command, operations, planning, logistics, and finance/administration. These components are discussed in Chapter 3 along with the specifics for the ICS. Many in the emergency preparedness field have modified Thomas P. "Tip" O'Neill's famous quote, "All politics is local" (O'Neill & Hymel, 1995) to "All disasters are local." Even though the local contingent typically addresses the initial response phase, response and eventually recovery assistance is needed at multiple levels including state and federal. In Chapter 3, these levels are discussed along with the importance of communication and the steps to take when the crisis hits. Details are also provided on the programs available to assist and support utilities in achieving a timely and effective response to emergency situations.

1.3.3 Recovery

Once the immediate response phase has been addressed, recovery must begin. In Chapter 4, the recovery phase is examined, including preparation and planning activities, managing and controlling the situation, financial issues, maintaining services, and repair and replacement. The recovery phase is sometimes difficult to distinguish from the response phase and in many cases, the two may overlap as critical information is collected over the course of the crisis.

1.3.4 Case Studies

The case studies presented in Chapters 3, 4, and 5 give real-world examples of disaster situations faced by WRRFs around the world. These scenarios provide vital lessons-learned and give the perspective of those who have faced the crises for which many facilities have plans. This information can assist facilities in updating their current plans or provide a valuable outlook on situations not currently covered.

2.0 EMERGENCY SITUATIONS AND POTENTIAL EFFECTS ON WASTEWATER UTILITIES

This section provides an introduction to emergency situations, hazards, and disasters in relation to wastewater collection systems and WRRFs. Definitions and short examples of the various situations that may affect the wastewater industry are given, including infrastructure failures, natural disasters, and human-induced incidents.

2.1 Infrastructure Failure

Maintaining critical systems and facility operations during an emergency situation are the first priorities. If at any point critical systems become inoperable, contingency plans must be implemented to quickly restore operations.

2.1.1 Industrial Accidents

2.1.1.1 Conveyance System Failures

The full system of pipes and lift stations feeding the facility should be assessed to determine the potential hazards associated with breakage such as release of untreated wastewater. Asset management approaches that include preventive maintenance can go a long way to reduce the risks associated with pipe breaks; however, at a minimum, a detailed map of key system components along with provisions and equipment for making quick repair should be available. These breaks may be more common, for example, in older piping systems, high water table areas, and seismically active zones where temporary above-ground hoses provide a quick solution to reroute water distribution when necessary. Critical elements in a conveyance system, including large interceptors, high volume lift stations, inverted siphons, and so forth, should be identified; and the condition and likelihood of an incident occurring should be quantified. Modeling software with the capability to provide criticality analysis are used by WRRFs, which can be especially helpful in identifying critical infrastructure. Planning is required for emergency pump bypassing and pipeline replacement.

2.1.1.2 Fires

Fires can represent an immediate danger to the facility and employees, making the importance of good preparation and planning even more essential. The National Fire Protection Association developed an effective outline for countermeasures to be taken in the instance of a fire. Each facility should clearly delineate the responsibilities of employees and contractors on site. A plan should be in place for both the internal and external communications and alarm system in addition to the evacuation plan. The plan should describe signals to be used to begin evacuation, primary evacuation route, and alternate evacuation routes in cases where primary routes could be blocked by releases of hazardous materials, wastes, gases, or fires. Periodic drills should be conducted to evaluate the effectiveness of the plan (Colonna, 2002).

2.1.1.3 Hazardous Chemicals

The presence of hazardous chemicals on site, though inevitable in most cases, presents a potential danger during disaster or emergency situations. Depending on the situation, preparation processes can include securing containers and tanks, reducing inventory if vulnerable, or increasing inventory if there is concern related to delivery following an event. When considering the storage of hazardous gases such as chlorine or sulfur dioxide and flammable or combustible chemicals, security should also be taken into account to restrict access to only authorized personnel. Any hazardous chemicals should be clearly identified in the facility material and waste inventory including the location, sources, and quantities. Safety Data Sheets must be available for each chemical of concern to meet 29 CFR 1910.1200(9) Hazardous Communications Standard Requirements with the chemical properties outlined in the response plan for addressing prevention, containment, mitigation, cleanup, and disposal measures.

2.1.1.4 Radiation Release

Discharge of radioactive materials to municipal wastewater systems may result in the potential for radiation contamination of the influent and treatment works. The Nuclear Regulatory Commission estimates that of the more than 22,000 regulated users of Atomic Energy Act radioactive materials, approximately 9,000 users have the potential to release radioactive materials to sanitary sewer systems (U.S. Department of Energy [USDOE] & U.S. Environmental Protection Agency [U.S. EPA], 2005). The four radionuclides most frequently reported in sludge are iodine-131, radium-226, americium-241, and cesium-137 (USDOE & U.S. EPA, 2005).

2.1.2 Electrical Outages

Electrical outages can have significant effects on the operation and functionality of the WRRF and various facilities, including pumping stations. Backup electrical power is essential to allow utilities to maintain essential functions during the emergencies and is required in several areas.

2.1.2.1 Community

The loss of power in communities surrounding a WRRF can cause a broad spectrum of effects. Depending on the duration and cause of the outage, evacuation of surrounding populations that feed the facility can actually reduce flows. However, the loss of power in the community can also mean the loss of power to lift and pumping stations, resulting in difficulties transporting the wastewater to the facility for treatment and potentially causing sewer system backups to homes.

2.1.2.2 Facility

Electrical outages present significant challenges to the operation of collection systems and WRRFs. Of particular concern are the critical needs of the facility, including such systems as pumps and lift stations. Consideration must be given if generators are to be rotated to serve multiple lift stations. The Florida Rural Water Association has full plans available for implementing variable frequency drive pumps for smaller generator use and how best to plan for generator "hopping" http://www.frwa.net (University of Florida & Florida Water/Wastewater Agency Response Network, 2018). U.S. EPA recommends classifying critical, secondary, and noncritical power needs, then identifying voltage, phase configuration, and horsepower/amperage requirements. Additionally, electrical equipment starting power demands are typically two to three times higher than their running demands, which may dictate a larger generator (U.S. EPA, 2011).

2.1.3 Communications

2.1.3.1 Continuity Planning

The importance of communications during and following a disaster situation cannot be emphasized enough. Multiple modes of communication are in some cases necessary to overcome problems as a result of outage or overloaded systems. In addition, the Government Emergency Telecommunications Service is an emergency phone service provided by the National Communications System in the Department of Homeland Security (DHS). This service operates on a variety of major networks to provide emergency access and priority processing for both local and long distance communications using a personal identification number system. Electrical outages at the facility can wreak havoc on computer systems and remote operations. Most facilities have backup systems in place through an uninterruptible power supply or generator. However, maintenance on these systems must be kept up to date and sufficient current verified to ensure that data systems and field instrumentation are running correctly; otherwise, the instrumentation package could be compromised. A plan should be developed for the continued backup of data and redundancy within the system. The use standardized procedures during the training can enhance the knowledge gained.

2.1.3.2 Technological Failures

Technological failures can encompass not only communication issues, but also the failure of systems that control facility operations. The NIST Cybersecurity Framework provides guidance on cybersecurity management to critical infrastructure organizations such as wastewater utilities. There are five core elements of the framework: Identify, Protect, Detect, Respond, and Recover. The key activities for wastewater utilities are to: (a) Create a detailed inventory of their computer systems, together with the connections to those systems; (b) Establish good cyberhygiene such as keeping computer software up to date, protecting with antivirus software, and securing physical and electronic access to systems; (c) Monitor their operations to identify potential cybersecurity incidents; (d) Prepare for a cybersecurity incident with backups and spare equipment so that operations can be restored quickly without resorting to payment of ransoms; and (e) Operate a recovery process that ensures the lessons are learned to better protect against the next attack.

In the planning phase of emergency response, it is important to consider redundancy in computer-based controls, particularly mechanisms for backing up records from critical systems including administrative data such as payroll. The supervisory control and data acquisition (SCADA) system used by the facility in addition to the remote terminal units and/or programmable logic controllers have technological vulnerabilities to malware and cyberattacks. The failure of the SCADA system can require the manual operation of systems typically maintained by computer systems. Training for these types of operations is important because this is not part of the typical daily routine. In addition, antennas may be lost depending on the situation, and their replacement should be a high priority.

2.1.4 Supply Chain

The importance of the supply chain is related to the critical needs of the facility, whether it is chemicals or equipment. In an emergency, some routes of delivery may be affected and a plan should be in place for alternative resources such as railcar versus truck. In addition, the supplies required to maintain service for the duration of a disaster, including the recovery phase, should be included in the overall disaster plan. This includes assessing requirements for chemicals (i.e., polymer, chlorine) and physical needs (i.e., equipment for recovery efforts and repair and fuel for emergency generators).

2.1.5 Transportation Accidents

Transportation accidents may pose risks to wastewater treatment systems in situations where materials or fuels are released to sewers. In addition, accidents involving transportation of chemical and raw materials to a WRRF or untreated sludge or Class B biosolids from the facility may pose risks to the environment and the public. Alternatively, if transportation accidents result in significant material spills in an area that feeds to the collection system, there can be effects to the WRRF, particularly if it is located within sensitive biological systems. The U.S. Department of Transportation, Pipeline and Hazardous Materials Safety Administration has released its *2020 Emergency Response Guidebook* (PHMSA, 2020). This guide details the response procedures for first responders in a hazmat situation. In addition, some larger facilities have plans for transportation safety within the confines of the facility, including travel direction, parking, and stopping restrictions.

2.2 Natural Disasters

Natural disasters can inflict grievous damage on collection systems and WRRFs. Possible effects include contamination of water from storm runoff or pipe breaks, as well as physical damage to infrastructure. While the magnitude, location, and recurrence interval of some natural disasters can be predicted using statistical modeling (Fukushima et al., 2006), building resilience within communities is key to avoiding the need for disaster mitigation (Kapur & Smith, 2011). Understanding a community's resilience can help emergency managers prioritize repairs or retrofit the infrastructure, while large-scale updates to infrastructure offer a unique opportunity to increase capacity of existing systems (Grant et al., 2004).

2.2.1 Biological

Biological disasters can manifest in many forms; however, they all are caused by living organisms. Exposure to pathogenic bacteria and viruses can result in the spread of infectious diseases, which can lead to epidemics and pandemics. An epidemic can be defined as an abnormal increase in the number of infectious disease cases in a population, whereas a pandemic is an increase in disease cases over a large region. An increased presence of disease within a population can result in an increase of staff on sick leave and a decrease in wastewater treatment efficacy if the biological contaminant of concern is not endemic or susceptible to current treatment mechanisms. A population fighting a disease may lead to increased antivirals and antibiotics in raw or inadequately treated wastewater. This may hinder microbial functions in WRRFs, thereby decreasing effectiveness. Pathogenic organisms are not the only cause of a biological disaster, however. Another biological hazard is overpopulation. Plant and animal infestations can block and damage wastewater systems caused by the rapid reproduction of flora and fauna. A notable case in animal infestation is the zebra mussel (*Dreissena polymorpha*), which is typically found in the Great Lakes but has been seen as far south as the Mississippi River in New Orleans. Effects range from clogging source water transmission systems, valves, and screens to damaging centrifugal pumps (Spellman, 2009). Multiple studies have examined the efficacy of various chemical treatments for their removal (Harrington et al., 1997; Klerks & Fraleigh, 1991; Matisoff et al., 1996; Van Benschoten et al., 1993). However, all of these showed the dose and contact time required for mortality were too high to be feasible (Spellman, 2009).

2.2.2 Geophysical

2.2.2.1 Earthquakes

Earthquakes, which result from breaking and shifting of rock below the earth's surface, can cause large-scale damage to structures of all kinds. Not only can the seismic vibrations cause buildings and bridges to collapse, but piping and other structures within the wastewater infrastructure are also at risk of damage (Fukushima et al., 2006). Similarly, in natural waters, significant changes in the architecture of groundwater systems can cause modifications in discharge and turbidity, while seepage of gases and other chemicals from within rock can promote increases in concentration of harmful chemicals (U.S. Geological Survey [USGS], 2003a). In the wake of the 2010 Haitian earthquake, Port au Prince's waterworks was badly damaged, leaving 2 million residents without access to clean drinking water, which resulted in disease outbreaks and greater fatalities (Padgett, 2010). The lack of adequate septage infrastructure also contributed to the disease outbreaks such as cholera.

2.2.2.2 Volcanic Eruptions

Volcanic eruptions occur when a break in the earth's crust allows molten rock and hot gas to escape to the surface. The ensuing explosion presents many hazards, including risk of fire, ash cover, mudflows, poisonous gases, and flying debris (FEMA, 2020). The second-most deadly volcanic eruption of the 20th century was the Nevado del Ruiz, Colombia, event in 1985 in which 25,000 people lost their lives primarily because of mudflows (USGS, 2012a). Besides the fatalities and economic damage that can result from volcanic eruptions, the addition of volcanic ash or mudflows to a WRRF can have effects in influent water quality—turbidity, acidity, and sometimes changes to water chemistry—that may render typical treatment processes ineffective. Enclosed facilities, however, fare far better and typically suffer minimal effects (USGS, 2012b).

2.2.2.3 Landslides

The mass movement of material downslope is called a landslide. Landslides can occur on varying time scales because of varying mechanisms, including overloading, slope instability, and water saturation of the material (FEMA, 2020). When this debris enters water systems, consumers may see an increase in turbidity and a decrease in the availability and quality of water (Bay Area Open Space Council, 2011). Most direct effects from landslides have been noted in the drinking water supply; the 1991 Antofagasta debris flow in Chile resulted in heavy damages to the system, leaving many without water service (USGS, 2012c). Using flexible pipefittings may help to avoid gas or water leaks, as some flexible fittings are more resistant to breakage.

2.2.2.4 Avalanches

Avalanches occur when a mass of snow moves downslope. Avalanches are typically triggered by rain, vibration, snowmobiles, or skiers that overload the slope causing a mass of snow to shift. Avalanches occur most frequently on slopes 35 to 50 degrees, which presents too great an angle of repose (Schweizer, 2004). Direct effects are similar to those of landslides, but avalanche control measures that use explosives may also increase organic residues in the watershed (USGS, 2003b).

2.2.3 Hydrological

Hydrological disasters, including rain, tsunamis, or severe storm events, are among the most common natural disasters in the United States. Damage from these events may include physical concerns like conduit failure, floating manhole covers, and inundation of infrastructure assets with water, mud, and debris. Biological pollutants may also become a concern, and municipal water supplies may become contaminated.

2.2.3.1 Floods

According to FEMA (2020), floods are one of the most common hazards in the United States. The 30-year national average for flood deaths is 127, with more than 10,000 lives claimed since 1900 (NOAA, 2012b). Floods may occur quickly or slowly, locally and regionally. Flash floods and dam breaks happen quickly, while a heavy rain might take time to saturate an area with water. Similarly, flooding may occur on any scale from a parking lot puddle to significant rain events like Hurricane Isaac in 2012 (Englande, 2008; FEMA, 2020). Inflow and infiltration can significantly increase the influent flow levels and result in treatment effects, particularly in biological systems. In addition, outfall locations may be inundated resulting in a backup of the system. Water quality is at great risk during floods as both physical damage and additional contamination are possible because the influx of spills (Cornell University, 2012).

2.2.3.2 Tsunamis and Tidal Waves

Englande (2008) notes the challenges faced by water quality specialists in the wake of tsunamis, tidal waves, and some flood events (especially those resulting from hurricanes) are nearly identical in damage and response to that of flood events from heavy rains or hurricanes. Although the causes differ between meteorological floods and tsunamis, the results are similar to that of storm surges or flash floods. Water quality issues arise from both contamination and physical damage to pipes and WRRFs (Brix et al., 2007; Cornell University, 2012). For example, the Tohoku earthquake in Japan in March 2011, which caused a tsunami and a subsequent nuclear reactor meltdown at Fukushima, killed 15,844 people, directly affected 41 million people, and resulted in $235 billion in damage (Davis et al., 2012).

2.2.3.3 Mudslides

A fluid type of mass wasting—like avalanches and landslides—mudslides typically occur in water-saturated soil on steep hillsides (USGS, 2012c), but can also result from volcanic activity. This rapidly moving debris can increase turbidity significantly and decrease the lifespan of screens in the facility. Notably, mudflows—resulting from wet conditions coupled with vibrations from the explosions—were the primary cause of death in the aftermath of the Nevado del Ruiz eruption (USGS, 2012c).

2.2.4 Meteorological

Meteorological disasters are caused by extreme weather conditions including high winds, flooding, or extreme temperature shifts. Also included in this category are lightning strikes, which can cause significant effects to the operations of a facility.

2.2.4.1 Hurricanes

Hurricanes are storms that have a rotating ring of thunderstorms around a central, storm-free eye. Hurricane season, or the part of the year when a hurricane is most likely to occur, spans from June to November in the Atlantic. The classification scheme for hurricanes considers their wind speed, central pressure, and capacity for damage and is known as the Saffir–Simpson Scale (Table 1.2).

The Federal Emergency Management Agency (2020) notes that flooding is the primary concern, as hurricanes can produce both widespread heavy rains and storm surges. In addition, the secondary effects such as power outages can significantly affect recovery efforts. Several days after Hurricane Katrina made landfall on the U.S. Gulf Coast, the U.S. EPA estimated that 1220 drinking water systems and more than 200 wastewater treatment facilities were affected (Copeland, 2005).

2.2.4.2 Tornadoes

Tornadoes, rapidly moving, narrow, rotating columns of air as defined by NOAA (2012b), are nature's most violent storms (FEMA, 2020). The NWS (2012) notes that tornadoes were the most deadly hazard in 2011, with the number of deaths climbing from 45 in 2010 to 553, because of the cataclysmic tornadoes in Alabama and other southern states in 2011. Tornadoes may pose not only the threat of water contamination, but also the destruction of aboveground wastewater infrastructure (U.S. EPA, 2020b).

2.2.4.3 Lightning Strikes

Lightning strikes can have both direct and indirect effects on facility operations. Indirectly, power surges resulting from strikes near the facility can damage electrical systems including pumps and generators resulting in destruction of the units or costly repairs. For example, in North Carolina in July 2012, an overflow of over 190 m^3 (50 000 gal) of untreated wastewater was discharged to a nearby creek caused by lightning damage to the pump and generator controls. The city responded by increasing dissolved oxygen levels in the creek using hydrogen peroxide to protect aquatic life (Daily Dispatch, 2012). A similar situation occurred in South Carolina when a lightning strike damaged a critical point in the intake system and more than 1515 m^3 (400 000 gal) of untreated water was discharged to the nearby Intracoastal Waterway (Myrtle Beach Online, 2011).

TABLE 1.2 Saffir–Simpson Hurricane Wind Scale (NWS, n.d.)

Category	Sustained winds	Types of damage due to hurricane winds
1	74–95 mph 64–82 kt 119–153 km/h	Very dangerous winds will produce some damage: Well-constructed frame homes could have damage to roof, shingles, vinyl siding, and gutters. Large branches of trees will snap and shallowly rooted trees may topple. Extensive damage to power lines and poles may result in power outages that could last a few to several days.
2	96–110 mph 83–95 kt 154–177 km/h	Extremely dangerous winds will cause extensive damage: Well-constructed frame homes could sustain significant roof and siding damage. Many shallowly rooted trees will be snapped or uprooted and block numerous roads. Near-total power loss is expected with outages that could last from several days to weeks.
3 (major)	111–129 mph 96–112 kt 178–208 km/h	Devastating damage will occur: Well-built framed homes may incur significant damage or removal of roof decking and gable ends. Many trees will be snapped or uprooted, blocking numerous roads. Electricity and water will be unavailable for several days to weeks after the storm passes.
4 (major)	130–156 mph 113–136 kt 209–251 km/h	Catastrophic damage will occur: Well-built framed homes can sustain severe damage with loss of most of the roof structure and/or some exterior walls. Most trees will be snapped or uprooted and power poles downed. Fallen trees and power poles will isolate residential areas. Power outages will last weeks to possibly months. Most of the area will be uninhabitable for weeks or months.
5 (major)	157 mph or higher 137 kt or higher 252 km/h or higher	Catastrophic damage will occur: A high percentage of framed homes will be destroyed, with total roof failure and wall collapse. Fallen trees and power poles will isolate residential areas. Power outages will last for weeks to possibly months. Most of the area will be uninhabitable for weeks or months.

2.2.5 Climatological

While meteorological disasters include weather conditions over a short amount of time, climatological disasters are considered the result of long-lived, large-scale weather events. Common examples of climatological events include dust storms, extreme temperatures, wildfires, and drought. All these disasters pose potentially hazardous effects on the wastewater industry. All potential effects of a climatological event must be carefully evaluated. Different regions are prone to different climatological events, so each emergency response plan must be tailored specifically for each facility.

2.2.5.1 Dust Storms

Dust storms are created when strong winds or fronts move through an arid area and transport loose sediments from dry surfaces. Sediment particles can be carried over vast distances, which allow contaminants to be transported over regions, eventually settling within raw or treated wastewater. These storms can also be of concern because of the effects dust can have on mechanical equipment, including surface damage. Additionally, dust can combine with oil and other lubricants to disrupt machinery and ground out electrical circuits (NOAA, 2003). In areas prone to dust storms, dust control strategies should be used.

2.2.5.2 Extreme Temperatures

Wastewater treatment systems are sensitive to extreme temperatures in a variety of ways. In extreme cold, it is crucial to control the temperature of water during treatment to prevent freezing. On the other hand, extreme increases in temperature can affect reaction rates and efficiency of treatment processes. High temperatures can also adversely affect oxygen transfer to biological systems because of both higher wastewater temperature and reduced blower efficiency caused by lower air density. Although the bacteria are somewhat resilient, temperature shifts result in different bacteria survival (i.e., mesophilic vs. thermophilic).

2.2.5.3 Wildfires

Vegetated areas with hot, dry climates are at an increased risk to develop wildfires. These uncontrollable fires can be widespread and disastrous. Wildfires produce great amounts of ash, debris, and nitrate, which cause an increase in wastewater treatment costs because of dredging, removal, and more advanced chemical treatment. Physical facilities can also be at risk from direct structural damage during wildfires.

2.2.5.4 Drought

Drought can be defined as a prolonged period of dry weather leading to insufficient water and moisture in an area. Droughts affect vegetation, crops, animals, and wastewater treatment. Water resources recovery facilities will be receiving less water. The water being received, however, may contain a higher concentration of contaminants. This stronger influent will likely have greater pathogenicity so chemical and mechanical treatment processes must be adjusted accordingly. A severe drought may even necessitate WRRFs to divert treated water to reuse options such as irrigation.

2.3 Human-Induced Incidents

Human-made disasters, also known as anthropogenic disasters, are the result of human intent, error, negligence, or because of failed systems. These disasters can occur because of intentional or unintentional human factors. Typically, two categories of hazards, technological and sociological, are associated causes of these disasters (Coleman, 2006; Pidgeon & O'Leary, 2000).

Sociological hazards are the results of living around others. Examples include noise, overcrowding, traffic jams, isolation, lack of privacy, and crowds (McKenzie et al., 2005). The types of anthropogenic disasters caused by sociological hazards can be divided into the following categories: personnel issues, civil unrest, criminal activities, and terrorism.

2.3.1 Personnel Issues

According to the DHS, critical infrastructure are the assets, systems, and networks, whether physical or virtual, so vital that their incapacitation or destruction would have a debilitating effect on security, national economic security, public health or safety, or any combination thereof (DHS, 2012). Personnel issues are those related to the loss, incapacitation, and/or death of those individuals necessary to maintain the functional operation of critical infrastructure components. Another personnel issue relates to cybersecurity risks. There is the potential for unintentional actions causing a cybersecurity issue (e.g., a user falling for an email attack, called phishing); deliberate actions as a result of a dispute with the organization (e.g., wiping data from a system); or after being fired (e.g., disabling systems using credentials that have not been changed after they left). These issues equally apply to contractors or vendors who do work for the wastewater utility.

The loss of critical staff would be felt most acutely during the recovery period of a disaster. As the community members return to their homes and rebuilding begins, an operational wastewater collection system is critical to ensure waste is managed appropriately to avert potential associated disease outbreaks. Facilities should have in their disaster planning guides consideration of staffing needs to quickly inspect and return to operation collection and processing systems. This topic is covered in much more detail later in the book (Chapter 4).

Critical personnel may also be lost because of labor disputes in unionized locations. Such an event occurred on September 30, 2012, in Detroit, Michigan, when wastewater employees in the Locals 207 and 2920 of the American Federation of State, County and Municipal Employees, which represent 950 public works employees, went on strike (AlHajal, 2020). In addition to the loss of personnel, solids hauling trucks refused to cross the union picket line early in the strike and smokestacks idled, indicating a low rate of waste processing (Huffpost Detroit, 2012). Communities and municipalities should have contingency plans in place should strikes occur.

2.3.2 Civil Unrest

Civil unrest may result when groups or individuals within the population feeling, rightly or wrongly, that their needs or rights are not being met, either by the society or by the overriding political system. When this results in community disruption of a nature in which intervention is required to maintain public safety, it is termed a civil disturbance. Examples of civil disorder include, but are not necessarily limited to, illegal parades, sit-ins and other forms of obstructions, riots, sabotage, and other forms of crime. The acts are intended to be a demonstration to the public and the government, but can escalate into general chaos. Peaceful, informational picketing on public property is protected activity, and picketers have a legal right to talk to people.

2.3.3 Criminal Activities

Crime is the breach of rules or laws for which some governing authority (via mechanisms such as legal systems) can ultimately prescribe a conviction. Individual human societies may each define crime and crimes differently, in different localities (state, local, international), at different time stages of the so-called "crime." According to the National Insurance Crime Bureau (NICB), there has been an 81% increase in claims related to theft of copper, bronze, brass, or aluminum from the 2006–2008 report to the latest 2009–2011 report (NICB, 2012). Of all the claims reported for the 2009–2011 report (25,083 claims), 96% pertained to the theft of copper. States with the highest number of claims include Ohio (2398), Texas (2023), Georgia (1481), California (1348), and Illinois (1284) (NICB, 2012). Water resources recovery facilities can be vulnerable because of remote sites such as pumping stations that are known to be unattended. The most common recommendations are camouflage, enclosure/locking, and replacing copper with other less valuable materials. Spartanburg County, South Carolina, has seen a 15% reduction in copper theft after legislation was passed requiring people who transport and sell nonferrous metal, which includes copper wire, copper-clad steel wire, copper pipe, copper bars, and catalytic converters, to obtain permits from the county where they plan to sell the metal (Arnold, 2012). Criminal activity in most cases is limited because of the large fixed structure nature of the equipment and is most often associated with petty crime by employees.

2.3.4 Terrorism

There is no single, universally accepted, definition of terrorism. The *Code of Federal Regulations* defines terrorism as "the unlawful use of force and violence against persons or property to intimidate or coerce a government, the civilian population, or any segment thereof, in furtherance of political or social objectives" (28 CFR Section 0.85). The Federal Bureau of Investigation further describes terrorism as either domestic or international, depending on the origin, base, and objectives of the terrorist organization. Cyberattack is another weapon that can be exploited by terrorists, because it can be relatively easy and low cost to deploy, and can be achieved remotely. Examples including deploying malware, disabling user access or view, or disabling control or safety systems.

Terrorists use numerous methods to achieve their objectives. These include the use of physical and cyberattacks and the use of specialized weapons such as chemical, biological, radiological, and nuclear agents. An example of a deliberate action from a disgruntled worker is the 2000 incident at Maroochy Water Services in Queensland, Australia.

Supervisory control and data acquisition systems are widely used to monitor and control operations in electrical power distribution facilities, oil and gas pipelines, water distribution systems, and WRRFs. The Maroochy system used two monitoring stations and three radio frequencies to control the operations of 142 wastewater pumping stations. A former contractor, motivated by revenge after he failed to secure a job with the Maroochy Shire Council, conducted the attack. Using a laptop computer and a radio transmitter, over a 3-month period, he released one million liters of untreated wastewater to a stormwater drain from where it flowed to local waterways (Slay & Miller, 2008).

Chlorine gas used at WRRFs may pose a threat to human life and property as a result of gas explosion or release. Presently, legislation by the U.S. EPA provide guidance in the storage of chlorine but there is no requirement with an associated punitive action in providing security for liquid chlorine. Many WRRFs have limited security, making these locations potential targets. Following the terrorist attacks of September 11, 2001, identification and protection of critical infrastructure have become national priorities. On October 10, 2001, the Subcommittee on Water Resources and Environment held a hearing on the security of infrastructure within the subcommittee's jurisdiction. What was discovered was that before September 11, 2001, wastewater infrastructure was not on the list of critical infrastructures, despite the fact that extensive networks of sanitary and storm sewers run near or beneath key buildings and roads, through the heart of business and financial districts, and everywhere in downtown areas of major cities.

Further action was desired to secure the infrastructure, and in 2002 the Wastewater Treatment Works Security Act was placed before Congress. Despite various iterations in 2005, 2007, and 2009, the bill has not yet passed. The bill attempts to amend the Federal Water Pollution Control Act (commonly known as the Clean Water Act) to require each owner or operator of a treatment works that stores or uses a substance of concern in quantities deemed by the administrator of the U.S. EPA to pose a security risk to conduct a vulnerability assessment of its treatment works, which shall include a site security plan to identify specific security enhancements. The act directs the administrator to develop and publish guidelines for carrying out a vulnerability assessment and, in developing such guidelines, to

1. Develop standards for assigning individual treatment works to four risk-based classifications taking into consideration a works' size and proximity to large population centers and the potential effects of intentional acts on public health or safety or the environment;
2. Designate any chemical substance as a substance of concern; and
3. Include a process for the revision, resubmission, and review of vulnerability assessments.

To aid the administrators, the U.S. EPA with assistance from the Departments of Veterans Affairs and Housing and Urban Development, in 2002 procured $500,000 through the Independent Agencies Appropriations Act (P.L. 107-73) to help fund the development of a vulnerability assessment tool for wastewater systems. The Vulnerability Self-Assessment Tool software is designed to provide wastewater utilities with the means to consider critical threats and identify key vulnerabilities to their systems (H.R. 5169—107th Congress, 2002).

In addition, the act directs the administrator to establish a threshold quantity for the release or theft of each such substance and sets forth requirements and deadlines for assessment certification, submission, review, approval, and disclosure. Authorization is given to the administrator to make grants to a state, municipality, or intermunicipal/interstate agency to conduct a vulnerability assessment of a publicly owned treatment works, to implement specified security enhancements to reduce identified vulnerabilities, and to provide for security-related training of treatment works employees and training for emergency response providers. Overall, the administrator is empowered

1. In coordination with the states, to provide technical guidance and assistance to small publicly owned treatment works on conducting a vulnerability assessment and implementation of security enhancements; and
2. To make grants to nonprofit organizations to assist in accomplishing purposes of this act (H.R. 2883—111th Congress, 2009).

Despite the lack of regulatory requirements, many communities have recognized this threat and as such have moved to alternative methods to disinfection such as the use of sodium hypochlorite, ozone, and other measures.

2.4 Multihazard Disasters

When assessing potential emergency situations, it is essential to remember that these seldom happen in isolation. One disaster can have many direct and indirect effects. Emergencies often occur in concert with cumulative effects that are greater than the sum of the individual events. For example, an area experiencing a drought may also experience increased influent contaminant concentrations. If the drought were accompanied by extreme heat, adverse effects on reaction rates in treatment processes may result. These are separate hazards whose effect needs to be assessed individually and aggregately to determine all potential risks and hazards. In preparing for disasters, WRRF operators must acknowledge that they do not function in a vacuum and that the functioning of other facilities in the region has an effect on them. It is important to realize what the role of the WRRF will be in a widespread disaster. This could include effluent distribution for irrigation purposes in times of extreme drought. Disasters rarely bring on a single hazard. Recognizing the multiple hazards of a disaster is crucial to understanding and preparing for potential effects.

3.0 DISASTER PLANNING AND RESPONSE

The National Academy of Sciences (NAS) *Disaster Resilience: A National Imperative* (2020) concludes that enhancing resilience is one way to reduce the effects of disasters. Chapter 5 outlines key aspects of resilience in disaster management and provide some examples from Australia. Resilience, the ability to adopt, absorb, recover from adverse events, is a critical and malleable attribute of physical infrastructure, government, communities, organizations, and individuals. Enhancing resilience is a fundamental principle of sustainable development that leads to improvements in the quality of life for communities and the ability to reduce effects and better respond when disasters strike.

Building resilience in wastewater systems and utilities requires anticipation of potential hazards and threats, including those that may be caused by changes in climate. Sea level change is an issue that must be factored into resilience plans for utilities in coastal communities even with the uncertainty associated with predicting sea level change (NOAA, 2012a).

This section provides a brief overview of disaster preparedness, response, and recovery.

3.1 Preparedness

Disasters strike unexpectedly with durations that in many cases are measured in minutes or hours. However, consequences of disasters persist for months and recovery may take years or decades (Figure 1.2). The preparedness phase of disaster planning is a continual activity that requires vigilance and regular and frequent testing, reassessment, and refinement of disaster management plans.

According to the World Bank (2010) disaster risk is composed of four elements:

1. Hazard—the likelihood and characteristics of the occurrence of a natural process or phenomenon that can produce damaging effects on a community;

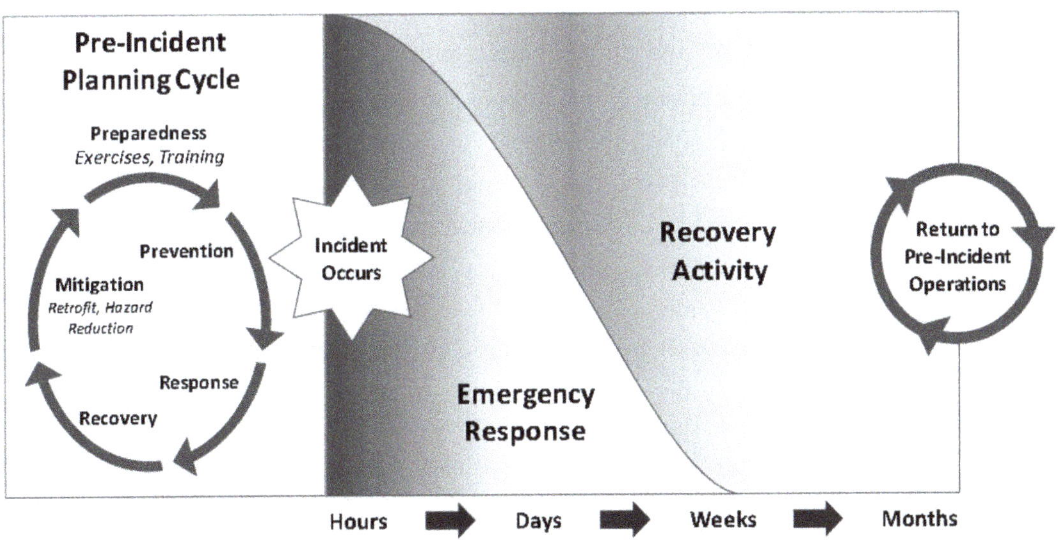

FIGURE 1.2 Planning, Response, and Recovery Phases (CIPAC, 2009)

2. Exposure—community's assets (people, property, and infrastructure) subject to the hazard's damaging effects;
3. Vulnerability—the potential for harm to physical assets, social capital, and political access; and
4. Consequences—the immediate and long-term results of the event affecting the exposure in a region or community.

The Critical Infrastructure Partnership Advisory Council (CIPAC) developed the *All-Hazard Consequence Management Planning for the Water Sector—Preparedness, Emergency Response, and Recovery* document to help water utilities incorporate consequence management concepts into their existing emergency preparedness, response, and recovery planning (CIPAC, 2009). The document provides customizable lists of preparedness, response, and recovery actions, consequence-specific lists of actions for potential hazards, and information on the National Incident Management System and ICS.

Checklists addressing each of the following crucial preparedness elements are included in the document:

- Know the utility's hazards and consequences,
- Connect with the emergency management agency in the utility's area,
- Identify response roles and responsibilities in the utility's organization,
- Ensure resources to maintain minimum operations,
- Create backup plans for key functions and resources,
- Identify and safeguard key utility information for response and business continuity,
- Understand and plan for cost reimbursement procedures,
- Train and practice and improve the utility's emergency response and recovery plans over time,
- Establish protocols for communication with the public,
- Be prepared to continue operations during recovery, and
- Be prepared to support employees during an emergency.

The NAS (2020) points to the following universal steps to improve resilience, including

- Engaging the whole community in disaster policy-making and planning;
- Linking public and private infrastructure performance and interests to resilience goals;

- Improving public and private infrastructure and essential services (such as health and education);
- Communicating risks, connecting community networks, and promoting a culture of resilience;
- Organizing communities, neighborhoods, and families to prepare for disasters;
- Adopting sound land-use planning practices; and
- Adopting and enforcing building codes and standards appropriate to existing hazards.

Disaster preparedness includes consideration of tangible structural construction related infrastructure issues as well as nonstructural issues including zoning ordinances, economic incentives, and management issues. The NAS studies point to the improved resilience that has resulted in buildings and the building industry since the introduction of the U.S. Green Building Council's) Leadership in Energy and Environmental Design program in 2000. The recently released Envision infrastructure sustainability rating system developed the Institute for Sustainable Infrastructure and the Zofnass Program for Sustainable Infrastructure at Harvard University will improve the resilience of horizontal civil infrastructure as its use increases over the coming years.

According to NAS (2020) the "process of building disaster resilience requires continuous assessment, planning, and refinement by communities and all levels of government."

3.2 Response

The ISO 31000:2009 standard, Risk Management—Principles and Guidelines, provides principles and generic guidelines on developing emergency preparedness plans and processes and provides a framework for response approaches (Döbbeling, 2012).

The ISO 22320:2011 standard addresses Emergency Management—Requirements for Incident Response, Command and Control (Tisserand et al., 2012) and the key interactions that take place during responses to emergencies as illustrated in Figure 1.3.

The International Organization for Standardization (ISO) is currently developing a crisis management standard, which will describe the fundamentals of a crisis management system and include recommendations and examples of relevant national authorities' experience in crisis management. ISO 11830 will be the first ISO standard that deals with crisis management in water and wastewater utilities (Tisserand et al., 2012).

Communications is one of the most important elements that must be focused on throughout all disaster phases but it is especially crucial during

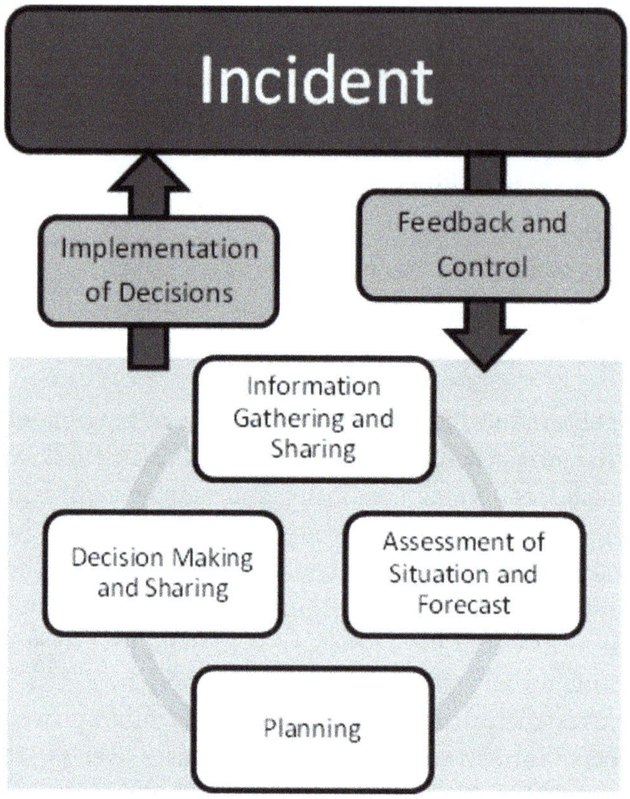

FIGURE 1.3 Response Activities

the response phase. The message map tool developed by risk communication expert Vincent Covello in the early 1990s was first widely adopted in the aftermath of the anthrax attacks of the fall of 2001 and has since become a key tool integrated into all sorts of disaster management plans (U.S. EPA, 2007). Message map templates (1) identifies the audience (stakeholders) for the message map and questions or concerns; (2) contains three key messages pertaining to the situation; and (3) the lists supporting information for the three key messages. The supporting information is blocked in groups of three under the key messages. Well-constructed message maps address the following goals:

- Educate and inform the public (stakeholders),
- Build and maintain trust and credibility between the general public and decision-makers during a crisis, and
- Create informed dialogue and decision-making among the public and figures of authority.

3.3 Recovery

Quick recovery from disaster should be a key goal for all wastewater utility managers and staff; however, the hard work of full recovery may take months to years. Wastewater utilities must be equipped to deal with the recovery phase and to coordinate their efforts with others affected by the disaster. Crucial to any recovery activity is learning from the disaster. Collaborative state-level water sector emergency response exercises conducted by the U.S. EPA from 2009 through 2011 (U.S. EPA, 2020a) pointed to the following key lessons learned through post-emergency assessments:

- Plan and coordinate with response partners before an incident.
- Be prepared to conduct damage assessments.
- Be prepared to request resources.
- Pre-identify backup power requirements.
- Plan for provision of alternative water supplies.
- Incorporate lessons learned into response plans.

4.0 SUMMARY

The NAS (2012) provides the following summary of issues related to disaster resiliency:

1. Disasters will continue to occur, whether natural or human-induced, in all parts of the country.
2. The population will continue to grow and age as will the number and size of communities; in some regions population decline and the number and size of communities will create a different set of challenges as tax bases decline.
3. Demographic data demonstrate that more people are moving to coastal and southern regions—areas with a high number of existing hazards such as droughts and hurricanes.
4. Public infrastructure is currently aging beyond acceptable design limits.
5. Infrastructure such as schools, public safety, and public health that are essential to communities are facing economically difficult times as the population grows and ages.
6. Economic and social systems are becoming increasingly interdependent and thus increasingly vulnerable should a key part of the system be disrupted.

7. Risk cannot be eliminated completely, so some residual risk will continue to exist and require management.
8. Effects of climate change and degradation of natural defenses such as coastal wetlands make the nation more vulnerable.

The World Bank (2010) points to the connection of three Cs: Cities, Climate, Catastrophes and the need for increased resilience required to address the increasing risks. The Rio+20 Issues Brief 8 (United Nations Conference on Sustainable Development, 2012) reinforces the need for disaster risk reduction and resilience building with the issues of the following three strategic goals:

1. Integrating disaster risk reduction in sustainable development policies and planning;
2. Developing and strengthening institutions, mechanisms, and capacities to build resilience to hazards; and
3. Systematically incorporating risk reduction approaches into the implementation of emergency preparedness, responses, and recovery programs

Reducing disaster risks to WRRFs contributes to community resilience and speedy recovery. To minimize vulnerability of infrastructure and/or communities, it is necessary to focus on resilience. Part of the resilience model concerns how people make sense of hazard effects and their consequences. Readjustment after catastrophic events requires the following factors:

- Coherence or comprehensibility (a way of making sense of the world);
- Perceived manageability of the situation; and
- Meaningfulness (understanding why the event occurred, its effect, and whether it signifies any value or purpose ultimately to the individual).

In technologically advised society new risks can be introduced every day. It is almost impossible to define all risks and hazards and effectively mitigate them. What is important is to empathize, enhance resilience, hidden abilities in human species to learn about surrounding environment and adopt.

5.0 REFERENCES

AlHajal, K. (2020, October 4). Detroit water workers vote to end strike after five days of picketing. *Michigan Live*. http://www.mlive.com/news/detroit/index.ssf/2012/10/detroit_water_workers_vote_to_1.html

Arnold, J. (2012, March 25). Stricter laws on scrap metal sales lead to fewer thefts. *Spartanburg Herald Journal.* http://www.goupstate.com/article/20120325/ARTICLES/203251017/1083/ARTICLES?p=1&tc=pg&tc=ar

Bay Area Open Space Council. (2011). *The conservation lands network: San Francisco Bay Area Upland Habitat goals and project report.* Bay Area Open Space Council.

Britton, N. R. (1986). Developing an understanding of disaster. *Journal of Sociology, 22,* 254–271.

Brix, H., Koottatep, T., & Laugesen, C. H. (2007). Wastewater treatment in tsunami affected areas of Thailand by constructed wetlands. *Water Science & Technology, 56,* 69–74.

Coleman, L. (2006). Frequency of man-made disasters in the 20th century. *Journal of Contingencies in Crisis Management, 14*(1), 3–11.

Colonna, G. (2002). *Introduction to employee fire and life safety.* National Fire Protection Association.

Copeland, C. (2005). *Hurricane-damaged drinking water and wastewater facilities: Impacts, needs, and response.* Congressional Research Service, Library of Congress. Order code: RS22285.

Cornell University—Cooperative Extension. (2012). *Water quality issues in emergencies.* http://waterquality.cce.cornell.edu/emergency.htm

Critical Infrastructure Partnership Advisory Council. (2009). *All-Hazard consequence management planning for the water sector: Preparedness, emergency response, and recovery.* Critical Infrastructure Partnership Advisory Council, Department of Homeland Security.

Cronin, S. (2018). *130.705 Emergency management study guide, part 2.* Institute of Natural Resources, Massey University.

Daily Dispatch. (2012, July 11). Wastewater overflows at Rowland's Pond due to lightning damage. *The Daily Dispatch*, Henderson, North Carolina. Retrieved September 2012 from http://www.hendersondispatch.com/view/full_story_free/19273785/article-Wastewater-overflows-at-Rowland%E2%80%99s-Pond-due-to-lightning-damage?instance=features

Davis, C., Keilis–Borok, V., Kossobokov, V., & Soloviev, A. (2012). Advance prediction of the March 11, 2011 Great East Japan Earthquake: A missed opportunity for disaster preparedness. *International Journal of Disaster Risk Reduction.* http://dx.doi.org/10.1016/j.ijdrr.2012.03.001

Department of Homeland Security. (2012). *Critical infrastructure.* http://www.dhs.gov/critical-infrastructure

Döbbeling, E. P. (2012). Emergency management: Global best practice for an incident response system. *ISO Focus, 3*(5), 28–31.

Englande, A. J. (2008). Katrina and the Thai tsunami—Water quality and public health aspects mitigation and research needs. *International Journal of Environmental Research and Public Health, 5*, 384–393.

Federal Emergency Management Agency. (2020). *Natural disasters*. http://www.ready.gov/natural-disasters

Fukushima, K., Kinoshita, K., Watanabe, T., & Masuta, T. (2006, October 21–25). Investigation and forecast earthquake damage for sewerage systems. *Proceedings of the Annual Water Environment Federation Technical Exhibition and Conference* [CD-ROM]; Dallas, Texas. Water Environment Federation.

Grant, G. M., Oram, P., Kimton, D., Roberts, L., & Beardsworth, M. (2004, October 2–6). Integration of Auckland's asset management and lifelines programs. Global proof of the benefits gained from combining asset management vulnerability assessments. *Proceedings of the Annual Water Environment Federation Technical Exhibition and Conference* [CD-ROM], New Orleans, Louisiana. Water Environment Federation.

H.R. 2883—111th Congress: Wastewater Treatment Works Security Act of 2009. (2009). In GovTrack.us (database of federal legislation). http://www.govtrack.us/congress/bills/111/hr2883

H.R. 5169—107th Congress: Wastewater Treatment Works Security Act of 2002. (2002). http://www.gpo.gov/fdsys/pkg/BILLS-107hr5169rh/pdf/BILLS-107hr5169rh.pdf

Harrington, D. K., Van Benschoten, J. E., Jensen, J. N., Lewis, D. P., & Neuhauser, E. F. (1997). Combined use of heat and oxidants for controlling adult zebra mussels. *Water Research, 31*, 2783–2791.

Huffpost Detroit. (2012, October 8). *Detroit Water Department strike ends, union says 36 fired workers will be reinstated*. http://www.huffingtonpost.com/2012/10/05/detroit-water-strike-union-ends-department_n_1943279.html

International Federation of Red Cross and Red Crescent Societies. (2006). *What is VCA? An Introduction to vulnerability and capacity assessment*. International Federation of Red Cross and Red Crescent Societies.

International Federation of Red Cross and Red Crescent Societies. (2018). *Types of disasters: Definition of hazard*. http://www.ifrc.org/en/what-we-do/disaster-management/about-disasters/definition-of-hazard/

Kapur, G. B., & Smith, J. P. (2011). *Emergency public health: Preparedness and response*. Jones and Bartlett.

Klerks, P. L., & Fraleigh, P. C. (1991). Controlling adult zebra mussels with oxidants. *Journal of the American Water Works Association, 83*, 92–100.

Matisoff, G., Brooks, G., & Bourland, B. I. (1996). Toxicity of chlorine dioxide to adult zebra mussels. *Journal of the American Water Works Association 88(8)*, 93–106.

McKenzie, J. F., Pinger, R. R., & Kotecki, J. E. (2005). *An introduction to community health* (5th ed.). Jones and Bartlett.

Myrtle Beach Online. (2011). Storm causes some issues. http://www.myrtlebeachonline.com/2011/06/20/2229891/in-our-towns.html

National Academy of Sciences. (2012). *Disaster resilience: A national imperative*. The National Academies Press.

National Insurance Crime Bureau. (2012, March 2). *Data analytics forecast report: Metal theft claims and questionable claims from January 1, 2009 to December 31, 2011*. [Technical Report] National Insurance Crime Bureau.

National Oceanic and Atmospheric Administration. (2003, April). Dust storms, sand storms and related NOAA activities in the Middle East. *NOAA Magazine*.

National Oceanic and Atmospheric Administration. (2012a). *Incorporating sea level change scenarios at the local level* [Technical Report]. NOAA National Coastal Services Center: Charleston.

National Oceanic and Atmospheric Administration. (2012b). *A severe weather primer*. www.nssl.noaa.gov/primer

National Weather Service. (n.d.). *Saffir-Simpson hurricane wind scale*. https://www.weather.gov/mfl/saffirsimpson#:~:text=The%20Saffir%2DSimpson%20Hurricane%20Wind,loss%20of%20life%20and%20damage

National Weather Service. (2012, May 8). *Summary of natural hazard statistics for 2011 in the United States* [Report]. NWS Office of Climate, Water and Weather Services and the National Climatic Data Center.

O'Neill, T., & Hymel, G. (1995). *All politics is local: And other rules of the game*. Adams Media Corporation.

Padgett, T. (2010). *The postquake water crisis: Getting seawater to the Haitians*. http://www.time.com/time/specials/packages/article/0,28804,1953379_1953494_1954584,00.html

Pidgeon, N., & O'Leary, M. (2000). Man-Made disasters: Why technology and organizations (sometimes) fail. *Safety Science, 34*(1–3), 15–30.

Pipeline and Hazardous Materials Safety Administration. (2020). *2020 emergency response guidebook: A guidebook for first responders during the initial phase of a dangerous goods/hazardous materials transportation incident*. Pipeline and Hazardous Materials Safety Administration, U.S. Department of Transportation.

Rodríguez, H., Quarantelli, E. L., & Dynes, R. R. (2007). *Handbook of disaster research*. Springer: New York.

Schweizer, J. (2004). Snow avalanches. *Water Resources Impact, 6*(1), 12–18.

Slay, J., & Miller, M. (2008). Lessons learned from the maroochy water breach. In E. Goetz, S. Shenoi (Eds.), *IFIP International Federation for Information Processing, Volume 253, pp. 73–82, Critical Infrastructure Protection*. Springer.

Smith, K., & Petley, D. N. (2008). *Environmental hazards: Assessing risk and reducing disaster* (5th ed.). Routledge.

Spellman, F. R. (2009). *Handbook of water and wastewater treatment plant operations*. CRC Press.

Tisserand, B., Sack, J., Zenz, T., & Ben-Ari, Y. (2012). Attention water utilities! Future ISO guidelines for crisis management. *ISO Focus, 3*(5), 13–15.

United Nations Conference on Sustainable Development. (2012, June 20–22). *Rio 2012 issues brief 8: Disaster risk reduction and resilience building*, Rio de Janeiro, Brazil. United Nations.

U.S. Department of Energy & U.S. Environmental Protection Agency. (2005). *ISCORS technical report 2004-04*; DOE/EH-0668; EPA 832-R-03-002B. U.S. Department of Energy; U.S. Environmental Protection Agency.

U.S. Environmental Protection Agency. (2007, March). *Effective risk and crisis communication during water security emergencies*; EPA Report 600/R-07/027. U.S. Environmental Protection Agency.

U.S. Environmental Protection Agency. (2011). *Is your water or wastewater system prepared? What you need to know about generators*. EPA 903-F-11-002. U.S. Environmental Protection Agency.

U.S. Environmental Protection Agency. (2020a). *Emergency/incident planning, response, and recovery*. http://water.epa.gov/infrastructure/watersecurity/emerplan/index.cfm

U.S. Environmental Protection Agency. (2020b). *Natural disasters*. http://www.epa.gov/naturaldisasters/

U.S. Geological Survey. (2000). *Landslide hazards* (USGS Fact Sheet FS-071-00). U.S. Geological Survey.

U.S. Geological Survey. (2003a). *Earthquakes–rattling the earth's plumbing system* (Fact Sheet 096-03). U.S. Geological Survey.

U.S. Geological Survey. (2003b). *Explosive-residue compounds resulting from snow avalanche control in the wasatch mountains of Utah* (Water-Resources Investigations Report 03-4007). U.S. Geological Survey.

U.S. Geological Survey. (2012a). *Nevado del Ruiz, Colombia.* http://vulcan.wr.usgs.gov/Volcanoes/Colombia/Ruiz/framework.html

U.S. Geological Survey. (2012b). *Volcanic ash: effects and mitigation strategies.* http://volcanoes.usgs.gov/ash/water/index.html

U.S. Geological Survey. (2012c). Worldwide overview of large landslides of the 20th and 21st centuries. http://landslides.usgs.gov/learning/majorls.php

University of Florida & Florida Water/Wastewater Agency Response Network. (2018, April). *Emergency response & preparedness best management practices for Florida's water & wastewater systems.* University of Florida.

Van Benschoten, J. E., Jensen, J. N., Brady, T. J., Lewis, D. P., Sferazza, J., & Neuhauser, E. F. (1993). Response of zebra mussel veligers to chemical oxidants. *Water Research, 27,* 575–582.

World Bank. (2010). *Natural hazards, unnatural disasters: The economics of effective prevention.* United Nations.

2

Emergency Planning

1.0 REGULATORY REQUIREMENTS 36
1.1 Emergency Planning Process 36
1.1.1 Why Plan for Emergencies? 36
1.1.2 Who Should Plan? 37
1.1.3 Resource Conservation and Recovery Act (Contingency Planning) 39
1.1.4 Emergency Plans Required by Regulations 39
1.1.5 Facility Response Plan 39
1.1.6 Spill Prevention, Containment, and Countermeasure Plans 41
1.1.7 Oil Removal Contingency Plan 42
1.1.8 Stormwater Pollution Prevention Plan 42
1.1.9 Resource Conservation and Recovery Act Contingency Plans 43
1.1.10 Cybersecurity 44
1.1.11 Risk Management Plan 44
1.1.12 Training and Community Awareness 45
1.1.13 Building Disaster-Resilience Community 45
1.2 Emergency Response Center and Operation 46
1.2.1 Introduction 46
1.2.2 Command and Control Structure 48
1.2.3 Incident Command System 49
1.2.4 Communication and Information Management 49
1.2.5 Decision Process 51
1.2.6 Organization Structure 52
1.2.7 Authority 52
1.2.8 Media Involvement 53
1.3 Risk Management Plans 54
1.3.1 Disaster Prevention and Mitigation Engineering 54
1.3.2 Communication and Consultation 55
1.3.3 Establish Context 57
1.3.4 Risk Identification 58
1.3.5 Risk Analysis 58
1.3.6 Risk Evaluation 59
1.3.7 Risk Treatment 59
1.3.8 Monitoring and Reviewing 61
1.3.9 Vulnerability Assessment 62
1.3.10 Risk Perception 66

1.4 Occupational Safety and Health Administration Emergency Action Plans 66
1.4.1 Introduction 66
1.4.2 Procedures for Reporting 66
1.4.3 Procedures for Emergency Evacuation 68
1.4.4 Procedures for Employees Who Remain to Operate 68
1.4.5 Employee Alarm System 69
1.4.6 Staff Training 69
1.5 Occupational Safety and Health Administration Fire Prevention Plans 70
1.5.1 Significant Fire Hazards 70
1.5.2 Procedures to Control Combustible Waste Materials 71
1.5.3 Procedures for Regular Maintenance of Heat-Producing Equipment 71
1.5.4 Staff Training 71
2.0 DEVELOPING AN EMERGENCY RESPONSE PLAN 72
2.1 Introduction 72
2.1.1 Emergency Planning Process 73
2.1.2 Elements of a Plan 74
2.1.3 Telecommunication Destruction 76
2.1.4 Resources and Integration of Agencies 76
2.1.5 Assessment of Community Resiliency 77
2.2 Exercising the Plan 77
2.2.1 Internal and External Emergency Planning 77
2.2.2 Planning Process 78
2.2.3 Exercises and Training 79
2.3 Revising the Plan 80
3.0 REFERENCES 80
4.0 SUGGESTED READINGS 81

1.0 REGULATORY REQUIREMENTS

1.1 Emergency Planning Process

1.1.1 Why Plan for Emergencies?

> "We are all part of the civil-protection process. Emergencies can be planned for by experts, but they will be experienced by the relief community and the general public alike. Therefore, we should all prepare for the next disaster as remarkably few of us will be able entirely to avoid it" (Alexander, 2002).

The chief motivation behind emergency planning is to reduce the likelihood of lives being lost. A second motivation is to reduce damage to the infrastructures. An emergency plan in its broad sense is to inform, instruct, and direct participants in measures needed to manage the direct and flow-on effects of emergency events that may lead to disasters. Who should advise on what actions are required, what resources to use (including human resources), expected likely scenarios, as well as desired outcomes or goals? Thus, a plan is also a means of coordinating and avoiding the duplication of effort and resources. By its very nature, an emergency plan must be devised, implemented, and tested, well before any emergency event is expected. Planning, therefore, needs to be undertaken for each of the four emergency management cycles—reduction, readiness, response, and recovery. Inefficiencies in planning lead to loss or damage that could have been averted. Thus, it is a moral responsibility for those involved with public or employee safety to ensure that emergency planning is carried out effectively and appropriately. Because it is difficult to predict when and where an emergency event will occur, the availability of effective, flexible, and practiced response and recovery plans is essential.

Emergency planning should be based on an all-hazards approach and should be as flexible and adaptable as possible to rapidly changing circumstances. Emergency planning includes

- Anticipating what might be required of relevant groups or organizations.
- Taking steps necessary to ensure the response is organized and coordinated.
- Taking into special consideration factors, such as authority and decision-making, prioritizing activities, and so forth.
- Planning for (1) role allocation, task performance, and integrated levels of planning; (2) reviewing and testing processes; (3) structures and response; and (4) developing strategies for successful operational emergency management.

1.1.2 Who Should Plan?

- Corporations
- National and local governments
- Families

Sound planning must come from federal, provincial, and local governments, regional councils, and corporations, particularly because they provide oversight for their communities:

- Encourage community awareness and participation; this would include their views and inputs.
- Facilitate planning among groups by providing training and educational opportunities.

- Encourage coordination and collaboration among the different organizations, emergency services, community, and cultural groups, including indigenous people.
- Set legal standards to govern other bodies in what is expected (the U.S. Resource Conservation and Recovery Act [1986] stipulates that local authorities keep and collect information about natural hazards in their communities). Ensure consideration of other nonstatutory documents prepared by local authorities and emergency services.
- Outline responsibilities of agencies.

Communities must prepare themselves by planning for emergencies, which planning, even at a basic level, can save lives and property. Many community service groups are already in place, including volunteer fire departments, service clubs, and church groups. Rural communities are often regarded as having inherently strong resilience and preparation for emergencies; however, given the distance over which many rural communities are spread—communications problems and low population—they are often less prepared than first appearance suggests.

In the best case, emergency for a family should include

- An emergency survival kit that is good for 3 days for each family member;
- Escape routines from the house and assembly areas;
- Knowledge of where and how to turn off gas, electricity, and water;
- Provision for pets, including food and confinement;
- Practice evacuations;
- Roles of each family member, including children;
- Adequate insurance and other protection measures;
- Alternative electric power, heating, and cooking sources;
- Additional clothing;
- Additional gasoline;
- Escape route out of town;
- Barter items;
- Personal protection; and
- Food and water supplies.

All members of the family, especially children, should be included in the creation of a disaster plan (Alexander, 2002, 2018). Often adults are concerned that discussing these issues with children will merely frighten them. But it is important to discuss with children the implication of fires, storms, and earthquakes, and the reasons for a disaster plan. This will serve to mentally prepare them in the event of a disaster. They will know what to expect and what are their responsibilities. Their ability to cope will be better than if they had not previously been instructed or involved.

1.1.3 Resource Conservation and Recovery Act (Contingency Planning)

The plan sets out an organized, planned, and coordinated course of action to be followed to minimize hazards to human health or the environment from fires, explosions, or unplanned sudden or nonsudden releases of hazardous waste or hazardous waste constituents.

Specific components of any contingency plan required by the regulation include

- Emergency procedures that facility personnel must take in response to fires, explosions, or hazardous waste releases;
- A coordinated emergency services plan agreed to by local departments, fire departments, hospitals, and state and local emergencies responses teams;
- A list of emergency coordinators that includes names, addresses, and telephone numbers of all persons qualified to act as emergency coordinators;
- A list of emergency equipment, including location, physical description, and capabilities of each item listed; and
- An evacuation plan for the safe egress of facility personnel.

1.1.4 Emergency Plans Required by Regulations

A number of emergency plans are required under various regulatory programs. Table 2.1 provides a list of these and presents the regulatory references for the circumstances in which they apply and the key elements to be included in each.

The following sections look at many of the listed plans and briefly describe the key elements, under what circumstances they apply, and whether they are applicable to water and wastewater utilities.

1.1.5 Facility Response Plan

Under 40 CFR Part 112.20, a Facility Response Plan (FRP) is required for the owner or operator of a nontransportation facility that, because of its location, could cause substantial harm to the environment by discharging

TABLE 2.1 Regulatory Emergency Plan Programs (RCRA = Resource and Recovery Act; OSHA = Occupational Safety and Health Organization)

Plan	Regulatory requirements
Facility Response Plan (FRP)	40 CFR 112.20
Spill Prevention Containment and Countermeasure Plan (SPCC)	40 CFR 112.3-12
State, Local, and Regional Oil Removal Contingency Plans	40 CFR 109.1-6
Stormwater Pollution Prevention Plan (SWPPP)	40 CFR 122.26
RCRA Contingency Plan	40 CFR 264.50-56
Risk Management Plan (RMP)	40 CFR 68.150-195
Process Safety Management (PSM) Plan	29 CFR 1910.119
OSHA Emergency Action Plan (EAP)	29 CFR 1910.38
OSHA Fire Prevention Plan	29 CFR 1910.39

oil or oil products to or on the navigable waters or adjoining shorelines. In addition, an FRP is required for a facility that handles, stores, or transports animal fats and vegetable oils. The FRP is also required for the transfer of more than 159 000 L (42 000 gal) of oil over water to and from vessels or has a total storage capacity of oil products exceeding 3 800 000 L (1 000 000 gal) under certain circumstances. A model FRP is included as Appendix F to 40 CFR Part 112. Key elements of an FRP include

- Emergency action response plan;
- Facility information;
- Emergency response information;
- Hazard evaluation;
- Discharge scenarios;
- Discharge detection systems used;
- Plan implementation;
- Self-inspection, drills, and training;
- Diagrams; and
- Facility security.

The FRP will likely not be applicable to a majority of water and wastewater utilities with the exception of some large facilities that use oil and oil products and are near sensitive waters or load or unload large quantities of oil or oil products and those facilities that are looking to treat animal fats or vegetable oils as a part of their energy recovery operations.

1.1.6 Spill Prevention, Containment, and Countermeasure Plans

Section 40 CFR Parts 112.1 through 112.12 provide the basic regulatory requirements for a spill prevention, containment, and countermeasure (SPCC) plan. The SPCC plan is required for any facility handling oils, including animal fats, vegetable oils, and greases. Part 112.1 requires an SPCC for all facilities with an underground storage capacity of more than 159 000 L (42 000 gal), or a facility with an aboveground capacity exceeding 5000 L (1320 gal) (counting only containers of 208 L [55 gal] or more). Facilities subject to 40 CFR Parts 280 or 281 are not covered by these regulations, but are subject to the *requirements* of Parts 280 and 281.

The regulations require that the SPCC plan have the full approval of management at the level necessary to commit resources to respond to a spill, and be prepared following good engineering practices. The SPCC plan should be certified by a professional engineer that it meets good engineering practices. The key elements of the plan include

- How the SPCC plan conforms to the regulatory requirements;
- A description of the physical layout of the facility, including a diagram;
- The type of oil stored on site and the size and material of containers in which it is stored;
- Discharge prevention measures;
- Discharge or drainage controls;
- Countermeasures to be taken for discharge discovery;
- Methods of disposal for recovered materials;
- A contact list with telephone numbers;
- The means for a person discovering the spill to relate information (unless covered by an FRP);
- Prior reportable spill events and response information;
- The prediction of direction, rate of flow, and total quantity that could result from equipment failures;
- List of containment or diversionary structures;
- Personnel, training, and discharge prevention procedures;
- Facility security elements;

- Loading and unloading drainage structures and procedures; and
- Inspection procedures.

The SPCC plan is the most likely oil-related plan to be required for water resource recovery facilities (WRRFs).

1.1.7 Oil Removal Contingency Plan

40 CFR Part 109 covers the requirements for state, local, and regional oil removal contingency plans. Although the water and wastewater utility is not required to develop such a plan, it may be subject to one or a party to one. The elements of such a plan include

- A list of authorities, responsibilities, and duties of all persons, organizations, or agencies;
- The establishment of notification procedures for early detection;
- The identification of critical water areas;
- Provisions for access to reliable communications systems;
- An established prearranged procedure for seeking assistance;
- Provisions to ensure that full resource capabilities are available to respond to an emergency;
- Development of agreements;
- Provisions for well-defined and specific actions to be taken after notification; and
- Procedures to facilitate recovery of damages and enforcement measures.

1.1.8 Stormwater Pollution Prevention Plan

Under 40 CFR Part 122.26, stormwater from certain industrial activities is required to be discharged under a National Pollution Discharge Elimination System (NPDES) permit. Covered also are treatment facilities larger than 3.8 ML/d (1 mgd) or those that are required to have an industrial pretreatment facility. All NPDES permits that include stormwater are required to have a stormwater pollution prevention plan (SWPPP). Regulations do not specifically detail the elements of an SWPPP, but the U.S. Environmental Protection Agency (U.S. EPA) has established the requirement that an SWPPP be added to NPDES permits wherever stormwater may be contaminated by hazardous materials. The key elements of an SWPPP are

- A detailed site map showing drainage areas and hazardous material storage areas,

- A materials inventory,
- Identification of past spills and leaks,
- The location of nonstorm discharges,
- Details of existing monitoring data,
- An evaluation of the site for potential spills and leaks,
- Best management practice procedures,
- Conducting an annual site compliance evaluation,
- Inspection procedures,
- Recordkeeping and internal reporting,
- Employee training, and
- Implementing appropriate controls.

Most water and wastewater utilities will be required to develop an SWPPP under their NPDES permits.

1.1.9 Resource Conservation and Recovery Act Contingency Plans

Under 40 CFR Parts 264.50 through 264.56, hazardous waste treatment, storage, and disposal facilities (TSDFs) are required to develop a contingency plan. There are few instances in which water and/or wastewater utilities would be defined as a TSDF. Water and wastewater utilities may find themselves defined as a hazardous waste generator, but not likely a TSDF. If the facility can be defined as a TSDF, it must have a Resource Recovery and Conservation Act (RCRA) permit issued by U.S. EPA or a designated state authority. If a facility has an SPCC plan, it is required to only add provisions covering hazardous waste management operations. If the facility is subject to TSDF rules, the contingency plan must include

- A description of the actions facility personnel are to take in response to fires, explosions, or unplanned sudden or nonsudden releases;
- A description of arrangements agreed to be local, fire, police, hospitals, and so on;
- A list of names, addresses, and telephone numbers of all persons qualified to act as the emergency coordinator;
- A list of all emergency equipment and decontamination equipment at the facility, their location, purpose of function, and a description of what they can do;
- An evacuation plan for facility personnel; and
- Detailed emergency procedures.

1.1.10 Cybersecurity

The NIST Cybersecurity Framework provides guidance on cybersecurity management to critical infrastructure organizations such as wastewater utilities. There are five core elements of the framework:

1. Identify,
2. Protect,
3. Detect,
4. Respond, and
5. Recover.

The key activities for wastewater utilities are to

a. Create a detailed inventory of their computer systems, together with the connections to those systems; Establish good cyber-hygiene such as keeping computer software up to date, protecting with antivirus software, and securing physical and electronic access to systems;
b. Monitor their operations to identify potential cybersecurity incidents;
c. Prepare for a cybersecurity incident with backups and spare equipment so that operations can be restored quickly without resorting to payment of ransoms;
d. Operate a recovery process that ensures the lessons are learned to better protect against the next attack.

1.1.11 Risk Management Plan

40 CFR Parts 68.150 to 68.195 under the chemical accident prevention program, require the development of a risk management plan (RMP). The regulations require that the RMP be submitted to U.S. EPA, certified by the owner/operator and updated at a minimum of every 5 years. Water and wastewater facilities that handle or store above a threshold quantity from a specific list of chemicals may be subject to RMP requirements. Chlorine gas used for disinfection is one of the covered materials, as is sulfur dioxide. The key elements of an RMP are

- An executive summary,
- A registration form,
- An offsite consequence analysis,
- A 5-year accident history,
- A prevention program depending on the type of process being used,

- An emergency response program, and
- A certification statement from the owner or operator.

Details of key elements can be found in the regulations. The owner or operator must update the RMP within 6 months of any changes to the covered processes at least once every 5 years, at least 3 years when a new process is started that has a new covered threshold chemical, and within 6 months of a revised process hazard analysis or offsite consequence analysis. If a covered facility removes a process from operation, such as switching to ultraviolet disinfection from chlorine gas disinfection, the facility shall submit a deregistration form. The RMP shall be corrected when a revised accident history is necessary or when there are emergency contact changes.

1.1.12 Training and Community Awareness

The most important principle of good disaster preparedness planning is that it must include training as a key component (Quarantelli, 1986). Cybersecurity awareness training is now a critical element of an organization's defense against a cyber incident. When it comes to cybersecurity, people are the weakest link in an organization, but they are also the most important protection. Teaching employees to identify phishing emails and to take care with good cyber-hygiene is a critical, yet easily overlooked activity.

The goal of training in emergency management is to bring order to chaos during a disaster. Other training opportunities to consider would include specific functions such as warning systems, evacuation routes and procedures, shelter locations, and damage evaluation procedures.

Training must be focused on a specific system the city is using to manage a disaster. For most cities, the incident command system (ICS) is used to manage disasters. The benefits of having ICS as a management resource are that it provides common language and terminologies among different departments and agencies. It includes a structure that can expand or retract in accordance with each specific situation, and it allows for other jurisdictions during the response and recovery phases of disaster. Training should also include key officials and must focus on the procedures that will take place in the emergency operations center. Other training opportunities include training for terrorism and specific functions such as warning, evacuation, sheltering, and damage.

1.1.13 Building Disaster-Resilience Community

There is no single definition of what "disaster-resilience" community actually means. Past experience and research indicate that individuals, groups and communities may each posses different degree of resilience which will vary over time. Many suggest a close relationship between resilient community and supporting resilient infrastructure. Also, Emergency Management Australia defines resiliency as "measure of how quickly a system recovers from failures." This definition is applicable to individuals, groups, community, and hard infrastructure. Cronin (2018) indicates that resilience comprises three components: dispositional, cognitive, and environmental. These features relate to the economic, emotional, and cultural aspects if recovery, respectively.

The concept of resilience possesses three key strength. First, the most common use of the term does not assume that disaster prevention is always possible, as does the disaster-resilient community concept. It recognises that natural hazard events will always occur and suggests the need for recovery efforts. A second strength of the term is that it captures the social variables that are neglected by the disaster-resistant community concept, one could assume there is no such thing "disaster-resistant community." Finally, the disaster-resilient community is a community, individuals, and groups, which constantly learn about the environment they live it. The environment means the natural environment (water feature, land, landscape), supporting infrastructure, hazards (natural and human-made).

Resilience, like resistance, is applies to natural hazards and may have a unclear relationships to technological and civil disaster. A process of assessment of resilience of community should be part of the emergency planning and training process instead the recovery process.

1.2 Emergency Response Center and Operation

1.2.1 Introduction

When an emergency occurs, diverse hazard effects often spread over a large area to create an operational management environment that transcends the expertise and jurisdiction of any one agency. Consequently, emergency response involves the coordinated and integrated activity of several organizations, many of whom have little or no contact with one another under typical circumstances. In an integrated response that is not planned for, and for which the necessary capabilities and competencies are not developed, the ensuing collaboration will reduce response effectiveness. Developing this capability or an integrated emergency management requires that attention is given to several issues.

Problems arise when a number of agencies are involved in a mass incident. These may include

- Competition for resources, which may be in limited supply;
- Differences in understanding terminology and concepts;

- Organizations viewing the situation from their own organizational "world view" and not considering the perspectives of others players;
- Difficulties with accessing information—its quality, meaningfulness, and interpretation;
- Communication breakdown, logistically and personally, within or between agencies; and
- Personal attitudes, values, perspectives, and assumptions that are consciously or unconsciously brought into the picture.

All these difficulties lead to divisiveness between agencies that are trying to achieve the greater goal: the alleviation of suffering, ensuring subsequent public safety, and limiting and repairing damage to infrastructure and lifelines. Breakdown in relationship external to an organization may also have internal repercussions, with increased stress and demands on those personnel.

To facilitate the work of various organizations, it is necessary to have some systems coordination in place. To eliminate or reduce divisiveness between agencies, an integrated approach is required. This approach first began in the United States after the California forest fires in the 1970s and became known as ICS.

Although the focus of integrated and coordinated management systems is on the response phase, functioning during this time improves drastically if precontact is achieved. This is best done incorporating the following practice:

- Planning meetings;
- Non-work-related events (sporting events, involvement in non-emergency community projects);
- An awareness of roles, responsibilities, and operational management of other organizations, and who holds those positions; and
- Established communication networks (ensuring, for example, that radio frequencies are congruent).

In practice, to develop integrated emergency management (IEM), several plans from different agencies, government and nongovernment, need to amalgamate, or at least mesh. This is the primary means of avoiding confusion and the doubling-up of resources. The following levels of planning typically exist:

1. A national plan to convert government policy into action:
   - Provide leadership and coordination for lower levels of government;

- Offer models for planning, mitigation, and training at lower levels; and
- Provide emergency resources where regional and local are insufficient.

2. Regional plans to provide coordination and continuity for local planning initiatives:
   - Offer models that can be developed at local level;
   - Identify regional threat, variations of hazards and vulnerability; and
   - Channel emergency resources wherever needed.
3. Organizational plans to provide for intraorganizational processes—individualized emergency plans to take care of details (schools, hospitals, transport systems).

Although plans at all these levels need to promote self-sufficiency, they also need to be developed to promote their integration at various levels. For example, this could be sharing the same emergency operations center or establishing lines of communication among disaster managers working under different plans.

1.2.2 Command and Control Structure

A traditional approach to managing large-scale emergency situations has been the military-style "command and control" method—a hierarchical, chain-of-command structure. The advantage of this structure was the clarity of who was in charge and held the chief decision-making role. This being the case, the time taken to make decisions and instigate implementation was relatively rapid. However, it is typically an unyielding and inflexible approach. Procedures were to be followed, hence adaptation to new information as the situation evolved was limited or possibly non-existent. There is also a danger in this respect with so much power lying on one person's shoulders. With that, at worst, comes the potential for abuse of the situation in terms of ensuring one's own agendas are met. At best, a narrower field-of-view and smaller range of skills, knowledge, and experience exist than if a more integrated team-structured, diversified approach were used. The emphasis on command and control is based more on the persons' ability to give out orders, rather than the implementation of flexible and effective interventions.

At the other end of the spectrum from the authoritarian-based method is the consensus approach in which consultation occurs. However, emergency situations do not allow time for long deliberations and democratic approaches to crisis management.

1.2.3 Incident Command System

The original ICS, developed for wildfires, has since evolved to an all-hazard system that uses a common organizational structure and standardized management principles. The system is built around a foundation of five significant components, shown in Figure 2.1. The system allows scaling, so for small incidents, all components could be managed by one person, an "incident commander," whereas each component could be managed separately for large-scale incidents.

The incident commander is in charge at the incident, and typically is the most senior person who arrives at the scene. The roles of this person include

- Establishing an incident command post,
- Assessing priorities,
- Protecting life and property,
- Controlling personnel and equipment resources through an appropriate organizational structure,
- Being accountable for responder and public safety,
- Acting as liasion to outside agencies and organizations,
- Developing and implementing an incident action plan,
- Coordinating overall emergency activities,
- Coordinating the activities of outside agencies,
- Authorizing the release of information to the media, and
- Keeping track of costs.

As incidents grow in size or complexity, a more highly qualified incident commander may step in to replace the initial responder.

1.2.4 Communication and Information Management

In an IEM environment, interagency and intraagency communication is essential for understanding complex, dynamic, and evolving emergencies, and for providing information for decision-making. These two issues are closely related. Decision effectiveness is a function of the capability to access data and render it meaningful within a time frame dictated by urgent demands. Crisis decision-making, however, also requires developing dedicated decision systems and capabilities.

It is not sufficient for any one agency to consider information and decision needs solely from its own perspective. In an IEM environment, the planning process must consider these issues from several perspectives. The ability of agencies to access and use scientific and other information is central to

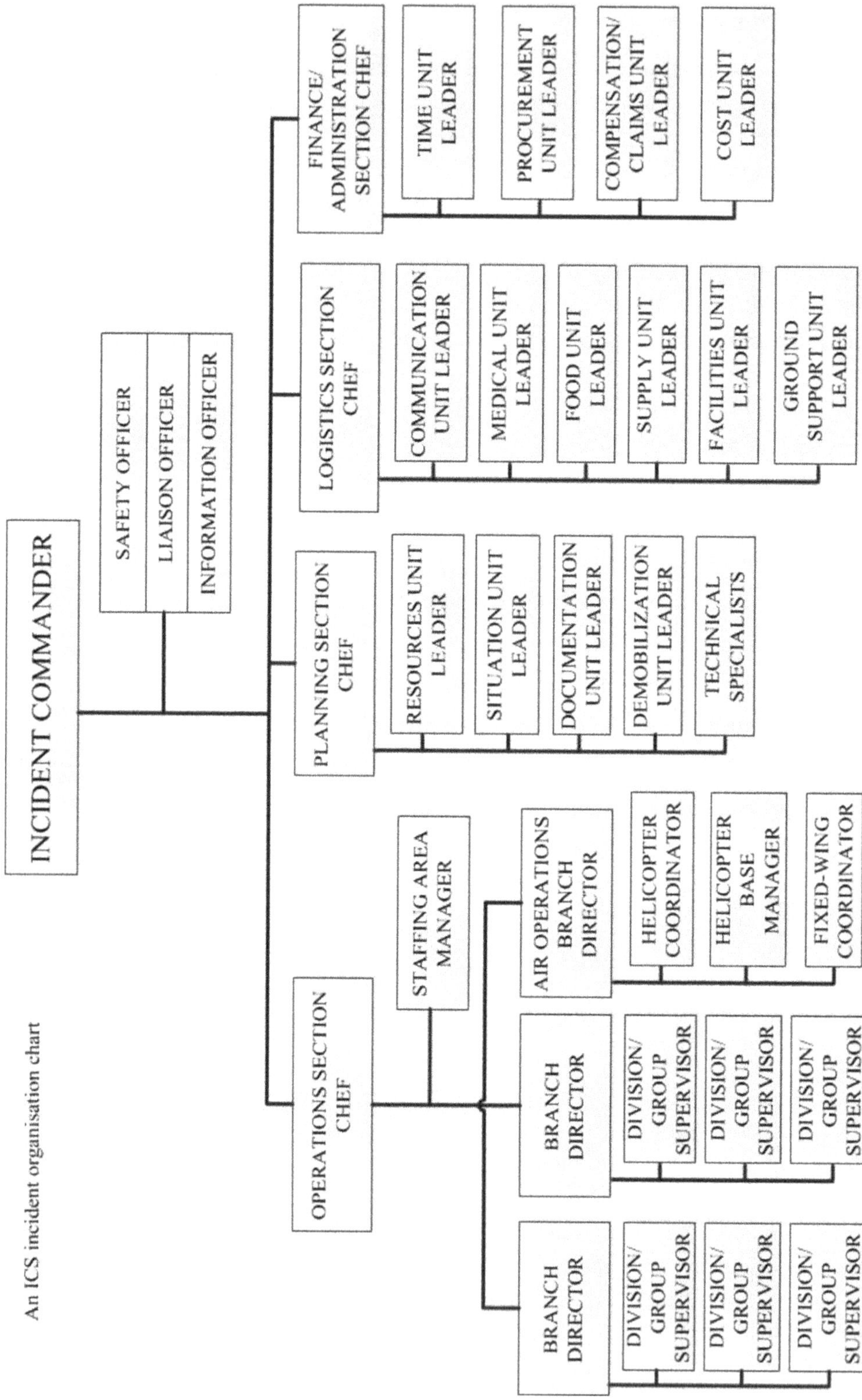

FIGURE 2.1 Incident Command Structure

hazard management. Although some communication problems emanate, for example, from hazard effects (such as communication infrastructure being affected by seismic activity), others can be attributed to the quality of crisis information management systems in conveying and understanding hazard consequences or the type of mitigation measures required. In doing so, emergency managers take scientific data (raw seismic data; eruption parameters, such as size and location; ash composition) and interpret and assimilate them to suit their specific needs.

The complexity of natural hazard phenomena and the diversity of agencies involved in an integrated response, make this a complex task. This task requires careful planning and development. Prominent issues in this context include: defining data and information needs; establishing information networks; decision-making; developing shared terminology, systems, and operating plans; and communicating with the public and the media.

1.2.5 Decision Process

The complexity of hazard effects, the uncertainty inherent in management data, and the diversity of agency information needs make information management a complex task, and one that requires comprehensive planning. Because developing the capability to access and comprehend information is an organizational task, it is equally important to be able to use this information to make response management decisions.

Emergency decisions are typically made against a backdrop of varying levels of complexity, time urgency, uncertainty, and shifting and competing demands. Being able to adapt to these changing circumstances becomes an important management capability. Acute stress can have positive or negative implications for emergency decision-making. In the short term, increased alertness, quicker reactions, and accelerated thinking combine to enhance performance and decision-making, particularly if individuals are trained to manage time pressures and to operate under conditions of high workloads and threatening conditions. If the level of demand increases, or the period of leadership responsibility is prolonged, performance and decision effectiveness decline. Typical problems under these circumstances include: "tunnel vision," failure to prioritize, "freezing," and loss of concentration. Thus, it is important to plan for the allocation and use of management resources to limit the risk of these adverse consequences. If, however, resources are insufficient to allow rotation of incident commanders, it is essential that those in leadership roles acknowledge the need to draw more on team resources under these circumstances.

As a consequence of the diversity inherent within the emergency operating environment, decision effectiveness is a function of the extent to which those involved possess a shared understanding of the response environment. It also requires that participants understand how their expertise contributes to different parts of the same plan while working toward common goals over time, including being able to anticipate the needs of those with whom they are collaborating.

The decision environment is also influenced by hazard characteristics. For example, managing earthquake hazards does not necessarily entail responding to an historical event.

1.2.6 Organization Structure

Response effectiveness is influenced by the nature of the organizational bureaucracy. Top-down predominantly autocratic procedures are ill-suited to emergency management in complex multijurisdictional settings. Internal organizational conflicts or issues can affect the ease with which multiagency integrated arrangements will be entered into and implemented effectively. A capacity for adopting a more responsive structure when managing crisis demands should not be assumed, but it can be developed through planning and exercise.

Emergency management systems, and the staff expertise to use them effectively, must be capable of dealing with sudden and radical changes in demands made on them, including

- Delegating authority (often with little or no warning);
- High demands on, and uncertainty in, information, communication, and decision processes;
- Transitorily adopting different operating and decision systems and processes;
- Working with organizations that use unfamiliar procedures and terminology;
- Managing the transitory transfer of control to external agencies when situation demands require it; and
- Coordination and assigning tasks in multijurisdictional or organizational operating contexts.

1.2.7 Authority

One of the main difficulties in multiagency efforts is the lack of certainty regarding who is in charge of the overall operation, if in fact there is such a person.

When agencies are not operating under an integrated banner, conflicts and power struggles can occur between those in charge of each agency. Issues between individual agency commanders are compounded through communication difficulties and terminology differences, resource competition, and overlapping goals. Having an integrated response system helps mitigate authority-based problems.

1.2.8 Media Involvement

Advances in communication technology increase the speed and vividness with which events can be reported. This has had a significant influence on the ability of media activities to contribute to staff and organizational problems. From a planning perspective, the manner in which the media operate can create problems for organizations faced with the task of managing a traumatic incident. Media representation of the event can also create difficulties. However, these problems are not inevitable, and the media play an important role. Collaborating with the media in the planning processes and defining their responsibilities pave the way for constructive and beneficial involvement during all phases of a disaster.

Positive outcomes typically reflect the ability of the mass media to distribute accurate information to a large number of people. Announcements that can be made available in this way include: information on what occurred; details of support services available and where they can be obtained; the need for evacuation, thereby, reducing the number of inquiries from community members; and assisting with the call-up of necessary personnel.

The media component of an emergency plan should cater for the following:

- Identifying a single media spokesperson from within the agency.
- Providing for press conference facilities in a location that prevents disruption to any part of the relief or recovery process.
- Arranging for information to be disseminated through daily press conferences and press releases only.
- Implementing a system to identify those with a legitimate right to be in a given location and to request information.
- Involving families and relatives of those killed or injured directly, through formal channels. This information should not be disclosed to anyone telephoning in (even if callers claim to be family or family representatives).
- Issuing detailed press releases on specific issues.
- Involving media in exercise and simulations.

Information commonly requested by media includes details about

- The number of people affected and the nature of their injuries;

- Property and environmental damage;
- Response, rescue, and relief activities; and
- Event characteristics.

1.3 Risk Management Plans

1.3.1 Disaster Prevention and Mitigation Engineering

Risk management plans should encompass all regulatory measures intended to shape the development and response to risks, both actual and potential. In this sense, risk management means a range of related activities for coping with risk; how risks are identified, evaluated, treated, monitored, and communicated; and how social perceptions and interventions to deal with risks are incorporated.

Two significant overriding elements in risk management are setting priorities and making choices. The way in which priorities are set depends on a number of external factors, including but not limited to, regulatory requirements, philosophical background, geographical locations, experience, and expectations. The overriding reason for setting priorities is that resources will never be available to treat all risks. The importance, then, of an effective risk management process is to provide an open and structured approach for maximizing the degree of risk reduction with available resources, including training and community awareness. This process also provides a robust and consistent basis for making and justifying decisions in risk management choices for emergency managers.

The risk management process is divided into a number of discrete interdependent components. The purpose behind each component is to facilitate courses of action that are technically appropriate and socially acceptable while working toward the overall goal of comprehensive risk management. Plans should run concurrently to maximize favorable consequences and minimize unfavorable ones. Recognizing and identifying each separate component as a distinct stage of the risk management process signifies the importance of the roles and responsibilities that are congruent to this type of process, further recognizing that expertise is required at every level of the decision-making process. Technical and analytical input from engineers, scientists, and health care professionals will be critical components to the risk management process.

A risk management plan consists of the following context: risk identification, risk analysis, risk evaluation, risk treatment; monitoring and review; and record of the risk management process. Figure 2.2 shows each step of a risk management process.

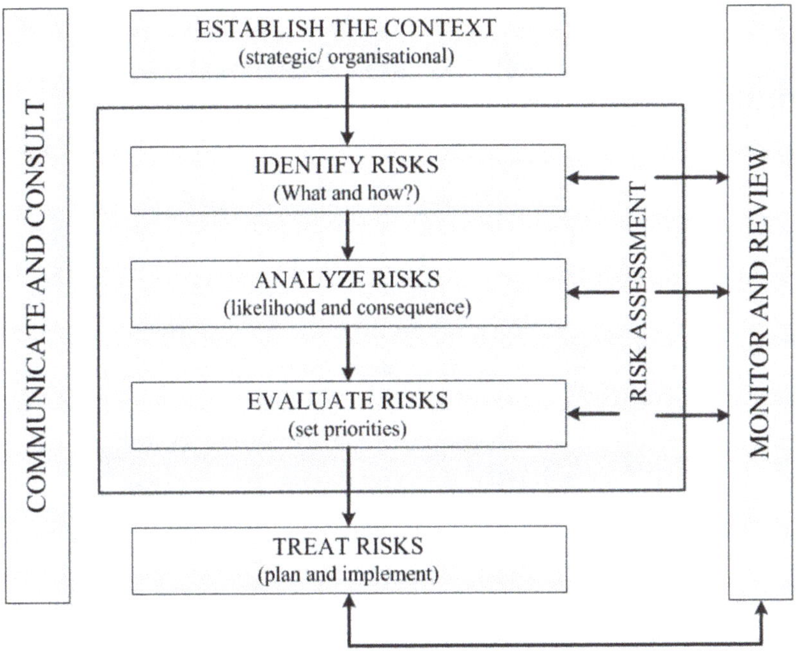

FIGURE 2.2 Risk Management Planning Process

1.3.2 Communication and Consultation

Successful risk communication requires its own specialized expertise and deliberate planning and evaluation. Senior managers need to devote attention and time to managing risk communication efforts—to enhance credibility, to ensure accuracy, to understand recipients and their concerns, and to gain the necessary insight into how messages are actually communicated.

Effective communication and consultation must also take place with the internal stakeholders. Involving staff at all levels seeking options and input, fosters a sense of coherence, ownership, belonging, and value. Table 2.2 is adapted from a report of the National Research Council's Committee on Risk Perception and Communication (Minister for the Environment, New Zealand, 2020). It comprises a risk information checklist of relevant topics for the design of a complete message.

Along with the checklist, the National Research Council's Committee on Risk Perception and Communication identifies four process objectives that it believes are key elements in improving risk communication:

- Goal setting—consideration of goals and constraints encourages realistic expectations, clarification of motives, and objectives.

TABLE 2.2 Risk Information Checklist

Information about the nature of the risk	What are the hazards of concern? What is the probability of exposure to each hazard? What is the distribution of exposure? What is the probability of each type of harm from a given exposure to each hazard? What are the sensitivities of different populations to each hazard? How do exposures interact with exposures from other hazards? What are the qualities of the hazard? What is the total population at risk?
Information about the nature of benefits	What are the benefits associated with the hazard? What is the probability that the projected benefit will actually follow the activity in question? What are the qualities of the benefits? Who benefits and in what ways? How many people benefit and how long do the benefits last? Which groups get a disproportionate share of the benefits? What is the total benefit?
Information and alternatives	What are the alternatives to the hazard in question? What is the effectiveness of each alternative? What are the risks and benefits of alternatives and how are they distributed?
Uncertainties in knowledge about risks	What are the weaknesses of available data? What are the assumptions on which estimates are based? How sensitive is the decision to changes in the estimates? What other risk and risk control assessments have been made and why are they different from those now being offered?
Information on management	Who is responsible for the decision? What issues have legal importance? What constrains the decision? What resources are available?

- Openness—two-way dialogue. The risk manager should ensure those in the organization understand what participants know and believe. They should also be aware and update what quantitative and qualitative information participants need to know to make decisions and how they think and conceptualize the risk.
- Balance—to ensure risk messages are not distorted or don't appear distorted. This could be carried out by a process of independent review.
- Competence—not only in the risk subject matter, but also in the techniques of risk communication.

1.3.3 Establish Context

This phase defines parameters and set the scene for the remainder of the risk management process and draws attention to the roles and responsibilities of various stakeholders and players. Four key factors are to be established:

1. The external or strategic context—this describes the external environment in which the organization operates, and the interface or relationship between the two. It includes
   - The business, social, cultural, financial, and political environment;
   - The organization's opportunities and threats;
   - External stakeholders;
   - Key business drivers; and
   - Legislation or other nominative prescriptions.
2. The internal context—this describes the internal environment of the organization. It includes
   - Culture;
   - Internal stakeholders;
   - Structures;
   - Resources—people, systems, capital;
   - Goals and objectives; and
   - The organization's strengths and weaknesses.
3. The risk criteria—criteria against which risk is to be evaluated (and later prioritized) are defined. Criteria could be based on operational, technical, financial, legal, social, environmental, or humanitarian considerations.
4. The risk management context—this refers to the goals, strategies, parameters, and scope of the activity, to which the risk management

process is being applied. Resources, costs and benefits, roles and responsibilities, and relation to other activities within the organization are defined.

Risk assessment includes three stages: risk identification, risk analysis, and risk treatment.

1.3.4 Risk Identification

This stage involves a comprehensive account of all possible risks, irrespective of their characteristics, to ensure that they are dealt with later in the process. The identification includes those risks that bring positive consequences, those that are beyond the control of the organization, and those that will later be deemed of little consequence. However, this process is best carried out in a systematic manner referring back to the elements described in the context, to ensure a complete and appropriate list. For each of the key elements, the process asked the following questions:

- What is the source of each risk?
- What might happen that could alter the achievement of the objectives?
- When, where, why, and how are these risks (positive and negative) likely to occur?
- Who might be involved or affected?
- What controls exist to treat specific risks?

To be meaningful across a broad range of agencies, the process of risk identification requires standardization of procedures.

1.3.5 Risk Analysis

Risk analysis takes a closer look at each risk with a view to developing a greater understanding to its sources, consequences (positive and negative), probability of occurrence, and existing controls. Similar risks can be grouped together, and those low-impact consequences may remain recorded, but not investigated or considered further. The analysis of risks helps to prioritize those that will require intervention, and ascertains the different options available. Analysis can be qualitative, semiquantitative, or fully quantitative.

The types of analysis used depends on details required, purpose of analysis, data, and resources available. Sources of information include past records and experience, literature and market research, experimental data, and expert opinion. Techniques for gaining understanding include structured interviews, questionnaires, multidisciplinary working groups, or the use of computer simulations.

Risk has two significant components: the likelihood or probability or occurrence and the magnitude of loss (or gain) (see Table 2.3). These are often combined as an estimate of the value of the risk, although complications arise because magnitudes of loss (or gain) are typically not commensurable. Estimates used in risk analysis will always contain some degree of imprecision or uncertainty. Hence, sensitivity analysis is a technique commonly used to test the effect of uncertainty in assumptions and the appropriateness and effectiveness of potential risk treatment control options.

1.3.6 Risk Evaluation

The risk evaluation phase is about making decisions with respect to future actions. It follows directly from the information ascertained in the risk analysis phase and weights it up with risk criteria established in the context phase. The purpose is to discover which risks need treatment and to prioritize these (refer to Table 2.4).

Risk evaluation estimates the acceptable or tolerable levels of risk—that is the degrees to which the organization, society, community, legislation, and others are willing to accept a certain level of risk. The simplest evaluation tools divide risk into three categories as shown in Figure 2.3.

1.3.7 Risk Treatment

Having evaluated all the identified risks, this phase investigates different treatment options, devises a treatment plan, and executes its implementation.

TABLE 2.3 Risk Analysis

	Consequences				
Likelihood	Insignificant 1	Minor 2	Moderate 3	Major 4	Catastrophic 5
A Almost certain	High	High	Extreme	Extreme	Extreme
B Likely	Moderate	High	High	Extreme	Extreme
C Possible	Low	Moderate	High	Extreme	Extreme
D Unlikely	Low	Low	Moderate	High	Extreme
E Rare	Low	Low	Moderate	High	High

TABLE 2.4 Qualitative Risk Evaluation

Risk				Refinement				Rating	Priority
	Likelihood	Effect	Level of risk	S	M	U	G		
Potato blight	C	4	Extreme	3	2	2	1	8	3
Carrot fly	B	4	Extreme	2	2	2	3	9	2
Locust plague	D	5	Extreme	3	3	2	3	10	1

According to AS/NZS 4360:2004 Guidelines a qualitative approach is suggested, where the CONSEQUENCES of risk impacts are rated on a 1 to 5 scale (with increasing severity) and the LIKEHOOD or recurrence of risk impacts are similarly rated on an A to E scale (with increasing likehood). The combination of these two factors enables a matrix to be constructed with which to categorize levels of risk from a variety of hazard types. However, there is no right or wrong way to assess the risk; it is the process the team of people should go through.

Likehood
A–Almost certain
B–Likely
C–Possible
D–Unlikely
E–Rare

Effect
1–Insignificant
2–Minor
3–Moderate
4–Major
5–Catastrophic

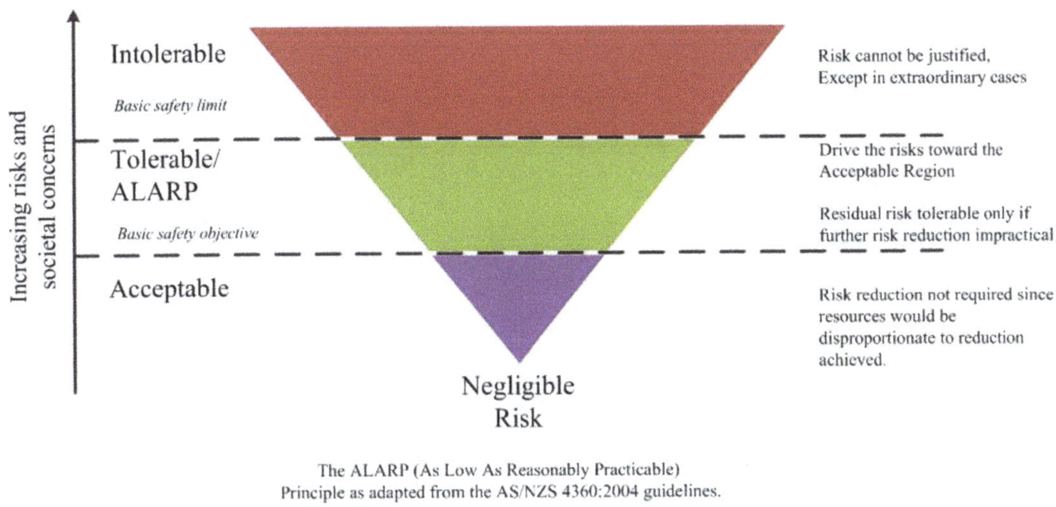

FIGURE 2.3 Risk Evaluation

Risk treatment options should be flexible and again, need to consider perceptions and values of stakeholders, community, and specialists, through consultation. Assessing risk treatment options involves evaluating cost, benefit, including those measures that are direct or indirect, tangible or intangible, and financially or nonfinancially measured. Cost–benefit analysis is an objective means of comparing the costs and benefits of the untreated risk with that of the treated risk. Typically, this is in monetary terms that are suitable for "hard," easily quantifiable costs.

In developing risk treatment options, new risks may be generated, which must then be incorporated in the established context of the framework, and undergo the subsequent analysis, evaluation, treatment, and monitoring considerations.

Planning for contingencies enables an organization to respond immediately with increased confidence and control while maintaining organizational stability and direction. This enables an organization to make use of unexpected advantages, or be resilient to unexpected disadvantages. Range of potential treatment options is included in Table 2.5.

An engineering-derived approach to risk treatment is demonstrated in Tables 2.6 to 2.8.

1.3.8 Monitoring and Reviewing

Ongoing review of each step ensures risk management plans remain relevant and acceptable in dynamic and changing environments. Performance indicators can be used to give a measure of the level of functioning and progress.

TABLE 2.5 Organizational and Functional Involvement in Risk Treatment

Reduction	Response
• Land use planning • Building code • Shelter criteria • Floodplain, tropical cyclone, storm surge, earthquake, hazardous materials spillages, mass transit accident, etc. • Management policies • Engineering works • Lifeline hardening • Regulatory provisions	• Forecasting • Warning • Operational deployment • Command and control • Reporting • Advising • Recording
Readiness	**Recovery**
• **Hazard** research and academic/science knowledge in natural and technological hazards areas • Town and project development planning • Information management • Public education • Allocation of lead and support agency roles and tasks • Planning • Exercises • Disaster management training • Resource identification	• Infrastructure restoration and recovery • Community restoration and recovery • Funding • Industry restoration and recovery • Environmental restoration

Third-party audits contribute a valuable independent perspective. Ongoing monitoring paves the way for continual improvement.

Documentation of assumptions, methods, data sources, analysis, results, and rationales behind decisions should be recorded at each step of the process. It is an indicator of good governance, reflects a methodical, reviewable, and auditable system, promoting the confidence of stakeholders.

1.3.9 Vulnerability Assessment

Vulnerability assessment of a community's infrastructure is a significant issue in hazard and risk management. Infrastructure, being the physical framework of a community, is often made up of several interrelated elements,

TABLE 2.6 Example of Generic Mitigation Measures

Preventative measures	Using hazardous zonation maps, carry out an audit of all infrastructure to identify elements at risk. Identify key components of each system, particularly in installations and buildings of post-disaster significance and infrastructure linking them. Nominate strategic components for upgrading or rerouting away from vulnerable areas. Upgrade all vulnerable strategic structures. Where it is possible to avoid hazard-prone areas, design, detail, or retrofit systems to minimize the effects of hazards and enable a rapid return to normal operation. Review design standards, specifications, and good practice guidelines, and revise them according to the priority of the components and level of finance available. Provide protection for sensitive plant and equipment (e.g., raise items above flood or surge level; fix and brace freestanding equipment in earthquake-prone locations; insulate computer hardware and control equipment from the effects of volcanic dust). Keep records on a database secure from damage, and ensure that maps of primary systems and district-by-district records of secondary systems are retained and accessible to users.
Spread the risk	Avoid dependence on single facilities and transport routines. Introduce redundancy or reinforcement into the distribution system for rerouting operations (advisable to safeguard supply during major repairs). Provide alternative sources of electricity supply.
Spread the responsibility	Widen ownership of the system, particularly for maintenance and operation, under regulatory control. Help inform communities to install and manage local systems, subject to regulations on minimum standards for security and quality of supply. Encourage user participation by promoting public–private partnership in community-based projects using forms of concession such as BOT (build/operate/transfer).

(continued)

together making up its so-called "lifelines." Reducing the vulnerability of the physical infrastructure (roads, WRRFs, power supply, telecommunications systems) to hazard will help reduce social and economic vulnerabilities within the community.

TABLE 2.6 Example of Generic Mitigation Measures (*Continued*)

Cover or minimize the effect	Provide insurance for physical losses, particularly for mechanical and electrical plant, to facilitate rapid recommissioning of the system. Encourage strategic users to install and regularly test standby power-generation equipment and ensure adequate fuel is available. Hold spare parts to replace critical items. Establish procedures for system failure and minimizing the effect of pollution.
Plan disaster management	Plan to minimize the time needed to return to normality. Arrange regular workshops and training programs for the continuing education of staff in hazard preparedness and reduction. Promote hazard awareness and the planning of facilities away from vulnerable areas.

TABLE 2.7 Risk Reduction Measures in Land Use and Planning

Preventative measures	Ensure that plans and designs for new development take account of any hazard of the area. Implement preventive or protective measures where practicable. Discourage development in hazard-prone areas where possible, and provide protection for vulnerable locations and facilities.
Spread the risk	Encourage decentralization of commercial and industrial sectors, while acknowledging transport implications. Develop alternative power suppliers. Ensure a range of suppliers and routes for delivery of resources.
Spread the responsibility	Encourage political decentralization. Encourage local communities within the city to manage the provision of infrastructure and services within their area.
Cover or minimize the impact	Plan evacuation routes and emergency vehicle access routes. Identify building for designing as emergency shelters.
Plan disaster management	Minimize the time taken to return to normality, and plan evacuation procedures.

TABLE 2.8 An Example of a Matrix for Scaling Consequences of Risks

	Consequence types					
Severity	Profit effect	Health and safety	Natural environment	Social/ cultural heritage	Community/ government	Legal
V	Increasing financial losses	Increases through pre-agreed measurement criteria or thresholds				
IV						
III						
II						
I						

1.3.10 Risk Perception

A fundamental issue in risk management is the problem of what to do when public perception and option differ from that of experts. Basic difference between risk assessment and risk perception can lead to conflict in risk management. Human perception is a filter through which hazards and risks are viewed and assimilated. Individuals carry their own set of values and beliefs based on knowledge, education, personality, and past experiences and interpretations. All these contribute to how an individual perceives situations of risks. Hence, not only does perception vary between experts and the public, but also between individual members of the public. For example, some people perceive themselves to be at high risk of getting cancer if they live near a chemical factory, whereas others in the same area think it unlikely. Perceptions we develop about what is safe are based just as much on social and psychological factors as they are on the actual physical characteristics of the hazard agent itself.

1.4 Occupational Safety and Health Administration Emergency Action Plans

1.4.1 Introduction

Occupation Safety and Health Association's (OSHA's) regulations for an emergency action plan are covered in OSHA standard 29 CFR 1910.38, *Employee Emergency Plans and Fire Prevention Plans*. OSHA states that the emergency action plan (EAP) covers the "designated actions that employers and employees must take to ensure safety from fire and other emergencies." OSHA requires the employer to record emergency action plans in writing unless there are 10 or fewer employees. If there are 10 or fewer employees, the employer may verbally communicate the plan. An EAP is a written document and the elements of this plan could typically include procedures for (1) emergency escape routes, (2) employees who remain to operate critical facility operations before they evacuate, (3) accounting for all employees after emergency evacuation has been completed, (4) rescue and medical duties for those employees who are to perform them, (5) means of reporting fires and other emergencies, and (6) contact persons for further information or explanation of duties under the plan. Figure 2.4 provides the Emergency Action Plan from OSHA 1910.38(a) standard (OSHA 3088, 2001).

1.4.2 Procedures for Reporting

The EAP should outline the preferred means of reporting fires and other emergencies. For example, depending on the facility, employees may dial

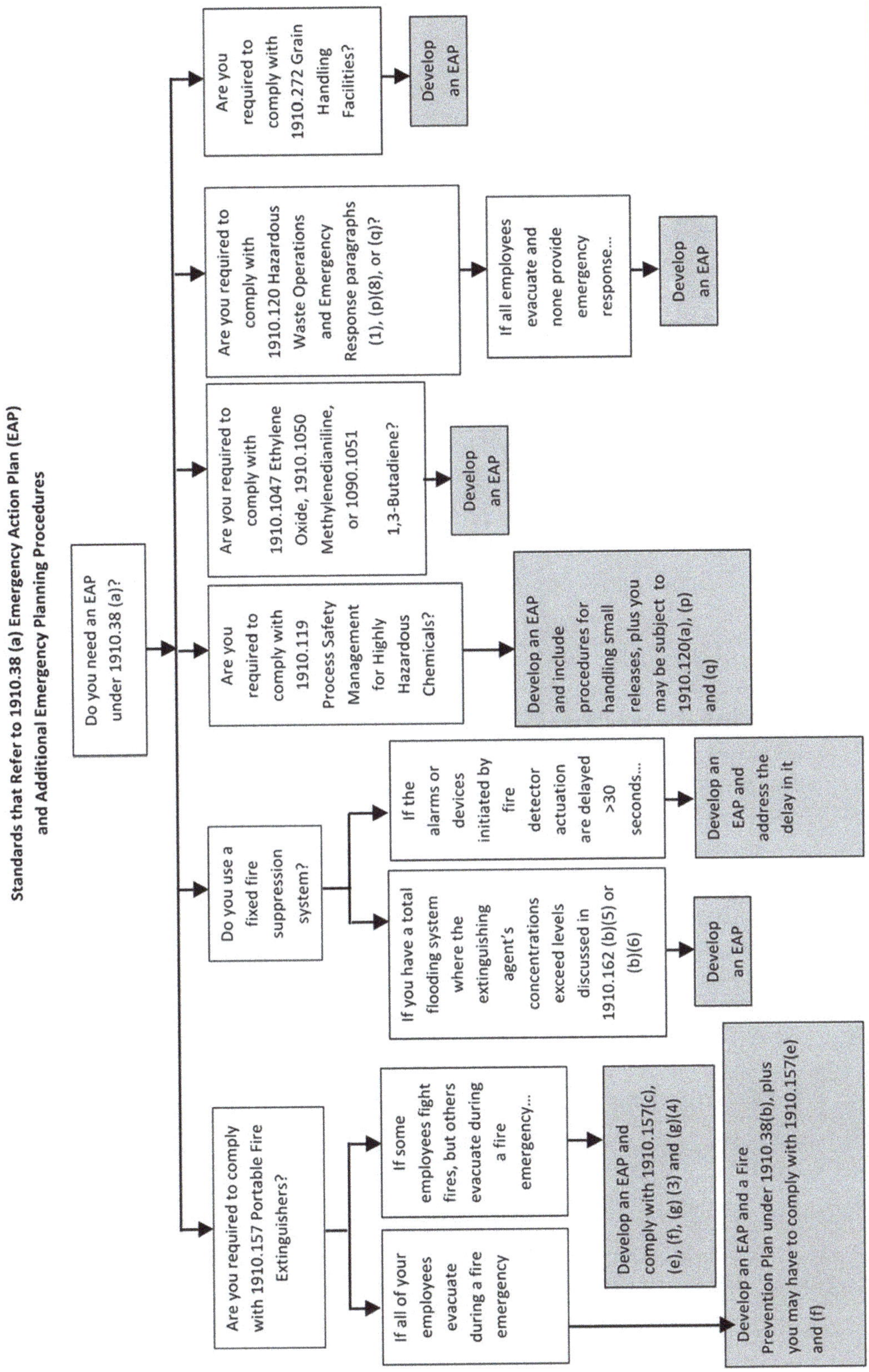

FIGURE 2.4 Emergency Action Plan from OSHA 1910.38(a) Standard (OSHA 3088, 2001)

911, dial an in-house emergency number, or pull a manual fire alarm. The EAP should include

- Addresses of nearest hospitals,
- Emergency telephone numbers,
- Fire emergency telephone numbers, and
- Names of emergency contact persons.

1.4.3 Procedures for Emergency Evacuation

Considerations for emergency evacuation procedures might include

- Specification of evacuation routes and exits that employees should follow during emergency conditions;
- Minimizing confusion, property damage, and employee injuries resulting from disorganized evacuation;
- Evacuation requirements for people with disabilities; and
- Employee awareness of what they need know and what is expected of them in all such emergency possibilities.

When the emergency is grave, total and immediate evacuation of all employees might be necessary; whereas in other situations, only a partial evacuation of nonessential employees might be needed with a delayed evacuation of others for continued facility operation.

The designation of refuge or safe areas for evacuation should be identified in the plan. Exterior refuge or safe areas may include parking lots, open fields, or streets that are located away from the site of the emergency and provide sufficient space to accommodate the employees. Considerations should be given to employees' transportation needs for communitywide evacuations.

1.4.4 Procedures for Employees Who Remain to Operate

- Depending on types of operations involved, sometimes certain employees might be required to remain behind at the facility to shut down special equipment before they evacuate building.
- Shutting down critical process equipment limits further damage to the equipment and reduces potential hazards, such as those from flowing liquids or gases under pressure.

- As an example, operations and procedures for personnel required at site for chemical spills emergency situations would typically include
  1. Confining hazard by closing doors and restricting access to the area,
  2. Shutting down or eliminating sources of ignition if spilled material is flammable liquid,
  3. Evacuating nonessential personnel from the area,
  4. Assisting the injured and keeping bystanders from entering the area,
  5. Reporting the spill to the emergency coordinator,
  6. Not reentering the building until it is safe, and
  7. Not attempting to clean up the site, unless trained.

1.4.5 Employee Alarm System

An employer must establish and maintain an employee alarm system. If the alarm system is used for alerting the fire brigade members (the in-house fire-fighting team) or for other purposes beyond notifying employees, then a distinctive signal should be used for each purpose and comply with the requirements in the OSHA 29 CFR1910.165 standard. The goal of this standard is to ensure that employees can be notified of the emergency. The method of transmitting the alarm should reflect the situation at the workplace. For example, a long horn blast followed by three short horn blasts could indicate an exterior fire emergency, and a siren could indicate a tornado or severe weather warning.

Some important points to consider for implementation of alarm systems might include (1) describing the alarm system to workers; (2) using alarm systems in the form of paging, audible tone systems, detectors, word-of-mouth, visual systems, tactile devices, vibration, air fans, for example; (3) using alarms capable of being perceived above ambient noise or light levels by all employees affected; (4) using tactile devices, vibration, or forced air to alert those who would not otherwise be able to recognize an audible or visual alarm; (5) using distinctive and recognizable alarms as signal to evacuate the work area or to perform actions designated under the EAP (i.e., allowing for reaction time for safe escape from the entire facility or work area, or for other emergency action); and (6) ensuring all emergency messages take priority when a communication system serves as the employee alarm system.

1.4.6 Staff Training

Education of staff is critical for successfully implementing the EAP. The employer shall provide adequate educational training programs to familiarize its employees with general principles of EAPs, including fire extinguisher use, hazards involved with incipient-stage firefighting, alarm systems, emergency exit routes, evacuation procedures, identification of employees to remain at site to shut down equipment, timely actions to be taken during emergency conditions, and so forth.

1.5 Occupational Safety and Health Administration Fire Prevention Plans

1.5.1 Significant Fire Hazards

The plan must address the significant workplace fire hazards at the facility and procedures for controlling the hazard in accordance with 29 CFR 1910.39(c)(1). The plan must be site specific and process specific. The following is a list of general fire hazards that could typically be found at WRRFs:

1. System hazards result from failures and misuse of electrical are
    - Equipment, including loose ground connections; wiring with frayed insulation; overloaded fuse, circuits, motors, or outlets and
    - Inadequate ventilation or improperly rated electrical equipment—which can lead to fires and explosions—in violation of *Standard for Fire Protection in Wastewater Treatment and Collection Facilities* (NFPA 820)(NFPA, 2020) requirements.
2. Chemicals and chemical gases present at a WRRF.
3. All maintenance activities, including but not limited to, cutting, welding, and open flame work.
4. All operational activities and processes, such as sludge driers and incinerators.
5. All office activities and practices, including trash and recycling practices; office electrical equipment practices with computers, printers, and other devices; other electrical practices, such as outlet usage and electrical extension cords; and designated smoking areas.
6. Flammable and combustible materials, which include materials used in operating and maintaining the facility. The flammable or combustible material is divided into two classifications:
    - Class A combustibles include material such as wood, paper, and others that can act as fuel and are typically found in areas such as offices.
    - Class B combustibles include flammable liquids, gases, and aerosols typically used in operating and maintaining the facility (oil,

gasoline, diesel, oxygen, acetylene, methanol, etc.). This can also include biosolids, screening, and grit.

For additional information, consult 29 CFR §1910.39(c)(1).

1.5.2 Procedures to Control Combustible Waste Materials

The following are typical guidelines and procedures. Procedures for each facility should be site specific and process specific. Typically, these include

- Dispose of trash daily;
- Dispose of biosolids, grit, and screening promptly, or place in adequate storage areas provided with proper ventilation, combustible monitoring, and other precautions; and
- Store and handle waste combustible materials in approved locations where vapors are prevented from reaching ignition sources, such as heating or electrical equipment, open flames, or sparks.

For additional information, consult 29 CFR 1910.39(c)(2).

1.5.3 Procedures for Regular Maintenance of Heat-Producing Equipment

These procedures should typically include the preventive and regular maintenance procedures recommended by the specific equipment manufacturer. This should include a minimum of daily inspection and continuous monitoring and/or alarming of the equipment to prevent overheating that could lead to a fire or explosion. Such monitoring may also include the equipment itself, as well as the feed material or fuel and the surrounding equipment (NFPA, 2020).

Personnel involved in the operation and maintenance of the equipment should be properly trained regarding the potential hazards of the equipment, including the potential of fire and explosions.

Consult 29 CFR 1910.39(c)(4) for further guidance.

1.5.4 Staff Training

Employees should be trained about fire hazards associated with materials, processes, and fire prevention equipment regarding the facility as a whole, and more specifically in the area to which they will typically be exposed to, operating, and/or maintaining. Such training shall occur

- At initial assignment, when newly hired or promoted, or at assignment change;

- At least annually; and
- When changes occur in equipment or process.

2.0 DEVELOPING AN EMERGENCY RESPONSE PLAN

2.1 Introduction

Good planning requires an understanding of demands likely to be made by emergencies and the variability of these demands both between disasters and over the course of the response and recovery phases. Plans that are sound, well-rehearsed, and regularly updated and maintained promote community preparedness and facilitate

- The prompt reinstatement of social infrastructure and community function; and
- The restoration of individual, group, and family functioning.

The existence of a plan does not guarantee these outcomes. Plans may exist but, if not rehearsed or based on appropriate assumptions about community vulnerability or capabilities, they will be less effective than anticipated. They may also contribute to the development of response-generated demands. Plan effectiveness can be diminished if they are based on the last experience rather than on promoting general preparedness through the use of the all-hazards approach. It is important to acknowledge that demands to be managed can emanate from preparation, hazard agents, and the emergency response.

1. Preparedness demands—preparedness demands can be described as follows:
   - Identifying hazard characteristics and their probable effects;
   - Developing hazard maps (for each hazard agent) indicating possible location, extend of likely effect, expected periodicity, damage/destruction expected;
   - Developing public education programs concerning hazard awareness and preparatory activities;
   - Compiling an inventory of community resources, including their location, state of readiness, quality, limitations; and
   - Developing and producing plans to integrate and coordinate a community with human and material resources, including voluntary and professional resources.

Plans covering the above issues need to be maintained and revised following testing in response to changes in knowledge and/or changes in community vulnerability.

1. Hazard effect demands—demands associated with hazard activity fall into eight categories:
   - Warning;
   - Preimpact preparations;
   - Search and rescues;
   - Physical, medical, and psychological care of the injured;
   - Care of the dead and attending to family needs;
   - Support and welfare needs;
   - Restoration of essential community services;
   - Protecting against continuing threats; and
   - Community order.
2. Response management demands—the activities and strategies implemented to deal with specific demands and problems generated by the hazard can generate their own demands. These fall under the following headings:
   - Continuing assessment of the disaster;
   - Communication, information management, and decision-making;
   - Mobilization and utilization of resources;
   - Coordination; and
   - Control and authority.

2.1.1 Emergency Planning Process

Planning needs to be a proactive and ongoing process. The starting point in this process involves defining organizational roles and objectives and defining the roles and responsibilities of the leadership. Responsibility should be vested in someone with the credibility and knowledge required to argue the case at senior levels.

The main steps of the emergency planning process are

- Deciding on policy,
- Assessing the nature of threats facing a community and who is threatened,
- Identifying resources,

- Selecting personnel,
- Training personnel,
- Developing contingency plans and emergency procedures,
- Locating and establishing an emergent operations center,
- Keeping the community informed,
- Revising the plans, and
- Practicing and rehearsing plans.

2.1.2 Elements of a Plan

The main elements of an emergency plan include

- A policy statement on the value of disaster planning. This should be prepared by a senior official. It will list the nature and level of commitment to activities undertaken and specify how and when testing and evaluation will take place. Any responsibilities will be identified by role or position rather than by identifying an individual.
- Legislative authority for plan development.
- Aims of the plan and conditions under which it comes into force.
- Identification of hazards and an assessment of the risk posed.
- Disaster scenarios.
- Relationship with levels of local and national government and emergency-related agencies.
- Authority organization chart.
- List of names, addresses, telephone and fax numbers, and email addresses of all relevant agencies. Developments on the Internet mean that this part of the plan can list URLs of source of information and assistance.
- Operations and warning systems:
    a. Types of warning,
    b. Distribution, and
    c. Obligations to receive warnings.
- Preimpact preparations:
    a. Relationship between hazard agent and preparations,
    b. Responsibilities of different agencies, and
    c. Location of greatest risk sites (physical, social, economic).
- Emergency evacuation procedures:
    a. Conditions under which evacuation is authorized;
    b. Routes to be followed and destinations; and

   c. Accommodating the special needs of the elderly, people with disabilities, the ill, and the institutionalized.

- Shelters:
   a. Locations and
   b. Facilities.
- Disaster control centers and subcenters:
   a. Location,
   b. Equipment,
   c. Operation,
   d. Staffing, and
   e. Interagency coordination.
- Communications:
   a. Between organizations and
   b. Within organization.
- Public information:
   a. Education and
   b. Communication with community.
- Search and rescue:
   a. Responsibility,
   b. Resources and equipment,
   c. Areas most likely to require serving, and
   d. Call-up and standby procedures.
- Community order.
- Medical facilities and morgues:
   a. Location,
   b. Transportation,
   c. Capacity, and
   d. Facilities.
- Restoration of community services:
   a. Lifelines,
   b. Order of priorities, and
   c. Responsibilities.
- Protection against continuing threat:
   a. Search for secondary threats and
   b. Actions to be taken if threats are discovered.

- Continuing assessment of total situation:
  a. Responsibilities and
  b. Distribution.
- Reciprocal agreements and links with other municipalities.
- Testing the plan:
  a. Disaster exercises and simulations,
  b. Exercise and simulation evaluation, and
  c. Feedback and review mechanisms,
- Revising and updating the plan, including consultation with all parties, and
- Plan distribution.

2.1.3 Telecommunication Destruction

Computer failures can also indirectly affect the community infrastructure. The following is an example of such a case. An explosion in 1976 in a Taft, Louisiana, chemical facility severely disrupted a state-of-the-art hazard monitoring system in the complex. The initial explosion was a minor incident, but occurred close to huge quantities of phosgene and chlorine gas. However, the plan operators had no idea that an explosion had occurred because their monitoring system collapsed. With the facility's monitoring system disabled, a new significant threat arose from poison gases and the perilous situation that created. It took hours before facility officials could clearly establish the actual threat. Meanwhile, there was significant social disruption of community life as thousands of people were evacuated in the middle of a rainy night.

2.1.4 Resources and Integration of Agencies

The diverse nature and unpredictability of emergencies make the process of planning difficult. There should be important bases covered: resources, integration of agencies, specific knowledge and skill, continuity, communication, comprehensive risk management, and approach.

The efficient planning and use of resources can be protecting lives and property, land use planning, and so forth. The knowledge of preevent community systems and needs can be anticipated during the planning process. Part of an effective planning process is identification of additional resources. This includes identification of an emergency operations center and its integration as an interorganizational resource.

Activities required during emergencies call for the services of several organizations and agencies. The common denominator between these agencies is that, ordinarily, the level of interaction between them will be minimal. Some organization will have existed before a disaster (emergency service organizations), but others may newly emerge (volunteer community groups). Planning is necessary for using all professional, established, and emergent volunteer groups. Involvement of the media in planning can emphasize the positive role they can play. Consultation with, and continued liaison among, these agencies with help to familiarize them with the systems used for control, information management, communication, and coordination.

Skills used in emergency management differ from those in routine organizational management. These require specialized training programs, regular exercises and simulations, and the formation and use of liaison and interorganizational communication channels.

Planning is not a one-time activity. It should be conceptualized as an evolutionary activity that is constantly revised and updated. Training and education of managers should also continue to be revised and updated.

Good communication is essential; analysis and soundness of technological and interpersonal aspects is a critical component. Dissemination of information about the planning process and activities should be made available to all community members, which is best accomplished by consultative mechanisms and enhancing community capability for self-care through education.

A comprehensive risk management approach is a necessary prerequisite for effective emergency planning. A well-designed emergency planning process will provide a structured, systematic basis for responding to unstructured, ambiguous events.

2.1.5 Assessment of Community Resiliency

People's resiliency is often determined by their values, attitudes, and practices and learned experiences. A resilient environment is one in which people and communities can maximize their potential for recovery and growth. The environment should be designed to allow the community and the individual to continue to function at an acceptable level and to mitigate and any adverse consequences, which water and wastewater infrastructure provide.

2.2 Exercising the Plan

2.2.1 Internal and External Emergency Planning

Internal and external emergency management plans have to be tested to ensure that they function properly when an accident happens.

The paper-plan syndrome refers to the existence of a written plan that is of no practical use in the real world, and in fact, may be detrimental to the preparedness of an organization. To avoid the paper-plan syndrome, a plan should

- Incorporate valid assumptions on how people behave and not how they are expected to behave.
- Involve users in the construction of the plan so there is some sense of ownership; this will aid to facilitate the plan.
- Be practical and usable.
- Be implemented in training exercises, so users are familiar with it and gaps in the plan can be addressed.
- Be general. Too much detail and the document is not user-friendly. Necessary details or specifics can be attached in appendices.
- Be integrated (interorganizational) and ensure each individual/organization realizes the importance of its part.
- Have an all-hazard plan.

2.2.2 Planning Process

The process of emergency management planning is more important than the plan itself. The process of planning is ongoing. It is evolving the documents to keep abreast of emergency-management developments, technology, additional involved agencies, legislation, and newly introduced hazards. Therefore, a plan is never finished; rather it is a work in progress (Cronin, 2006, 2018). According to (Cronin, 2006, 2018), there are four dangers in the recent completion of an emergency management plan:

- A feeling that the job is done. There can be a false sense of security attained in having a written document titled "emergency management plan" sitting on the shelf. It can give the impression that an individual or organization is prepared to meet disaster. However, without training, reviewing, and updating, the plan is almost useless.
- That it is produced for the wrong reasons. Often organizations must have a physical plan in place to be accredited, or it may be a legal requirement of a business regarding staff safety protocols. The incentive is to create the document by the due date to get the all-important legal nod or accreditation tick. Unfortunately, with this being the focus, the plan is not necessary in a viable form. It is looked on as an annoying chore, as opposed to an existing, proactive opportunity for positively addressing a company's needs.

- That it is produced as an end product.
- That it is unacceptable to the users. Being involved in planning brings the plan to life for those individuals who help create it. It makes it real, meaningful, and acceptable to users. It is not an untouchable document that has come down from on high or from outsiders who have no knowledge of how that organization or department really functions.

2.2.3 Exercises and Training

Exercises and training are the most common ways in which emergency plans are tested and enlivened after they are created. These are also the main means of training the actors within a plan of their roles and responsibilities. There are a number of ways training can take place.

1. Table-top exercises—these are valuable in that they are low cost, safe environments without harm to participants. A number of techniques can be used: small group work and role play and brainstorming sessions.
2. Simulated or field exercises—these take a great deal of organizing and involve considerable expense. They are invaluable for a number of reasons:
   - A more realistic interaction can be created between players, especially interorganizational relationships;
   - Roles and responsibilities among participants can be sorted out;
   - Participants' knowledge, skill, and confidence can be increased;
   - Gaps in resources or managerial processes and logistics can be recognized;
   - Otherwise unforeseen obstacles or difficulties can be recognized;
   - Alternative practices can be recognized;
   - The public's awareness can be heightened to the realities of disasters occurring locally;
   - They are one of the best ways to assess whether a plan will work efficiently;
   - They provide a safe environment, although dangers still exist;
   - The entire disaster scenario can be simulated, during which one or many components may be used; and
   - They can provide assurance that mistakes are acceptable and can be part of the learning process.

3. Overseas/national training experience at real-life emergencies—real-life experiences add to ability and knowledge, interest, and confidence in developing and implementing emergency plans, and lend a sense of participation and teamwork in a real-life event. They allow the evaluation of other systems, and give a firsthand impression of how factors, such as cooperation, function among volunteers, armed forces (under the guise of material law), medical and emergency services, and the public.

2.3 Revising the Plan

Plan revision and update is an essential part of emergency plan implementation. Organizations, policies, staff, legislation, environments, and almost every feature of the community will change, including social conditions for which a plan may be designed. Hence, the processes must be designed to update the plan. Six-month review and annual update of the plan are necessary.

As described above, the planning process is more important than the plan itself. The emergency management planning is an ongoing process. These reviews can be used also to revise the plans from field exercises or table-top scenarios in which participants are able to provide feedback in briefing sessions. In this way, a plan remains as a living document. There should never be dangers of ink drying on a good emergency management plan.

3.0 REFERENCES

Alexander, D. (2002). Nature's impartiality, man's inhumanity: Reflections on terrorism and world crisis in a context of historical disaster. *Disaster, 26*, 1–9.

Alexander, D. (2018). Nature's impartiality, man's inhumanity: Reflections on terrorism and world crisis in a context of historical disaster. *Disaster, 28*, 21–32.

Cronin, S. (2006). *130.705 Emergency management, study guide, part 2.* Institute of Natural Resources, Massey University: Palmerston North, New Zealand.

Cronin, S. (2018). *130.705 Emergency management, study guide, part 2.* Institute of Natural Resources, Massey University: Palmerston North, New Zealand.

Minister for the Environment. (2020). *Report of the national research council's committee on risk perception and communication.* Minister for the Environment. Wellington, New Zealand.

National Fire Protection Association. (2020). *Standard for fire protection in wastewater treatment and collection facilities*. NFPA 820. National Fire Protection Association.

Occupational Safety and Health Administration. (2001). *How to plan for workplace emergencies and evacuations*. http://www.osha.gov/Publications/osha3088.pdf

Quarantelli, E. L. (1986). *Research findings on organisation behaviour in disasters and their applicability in developing countries* (Preliminary paper #107). Disaster Research Center, University Delaware.

U.S. Resource Conservation and Recovery Act. (1986). http://www.epa.gov/lawsregs/laws/rcra.html

4.0 SUGGESTED READINGS

Adams, J. (1995). *Risk*. UCL Press.

Alexander, D. (1992). Natural disasters: A framework for research and teaching. *Disasters, 15*, 209–226.

Alexander, D. (2000). Scenario methodology for teaching principles of emergency management. *Disease Prevention and Management, 9*, 89–97.

Alexander, D. (2002). *Principles of emergency planning and management* (p. 340). Terra Publishing.

Anderson, M. B. (1994). Vulnerability to disaster and sustainable development: A general framework for assessing vulnerability. In M. Munasinghe & C. Clarke (Eds.), *Disaster Prevention for Sustainable Development: Economic and Policy Issues*. The International Decade for Natural Disaster Reduction and the World Bank.

Angus, D. (1997). Application of a risk management approach in a Queensland disaster management context. *Australian Journal of Emergency Management, 12*(1), 31–34.

Auf der Heide, E. (1989). Chapter 4–disasters are different. In *Disaster response*. C. V. Mosby.

Australian and New Zealand Risk Management Standards AS/NZS ISO 31000:2009.

Bachrach, K. M., & Zautra, A. J. (1985). Coping with a community stressor: The threat of a hazardous waste facility. *Journal of Health and Social Behavior, 26*, 127–141.

Barberi, F., & Carapezza, M. L. (1996). The problem of volcanic unrest: The Campl Flegrei case history. In R. Scarpa & R. I. Tilling (Eds.), *Monitoring and mitigation of volcanic hazards* (pp. 771–786). Springer Verlag.

Bates, F. L., & Peacock, W. G. (1987). Long term recovery. *International Journal of Mass Emergencies and Disasters, 7,* 349–365.

Begum, R. (1993). Women in environmental disasters: The 1991 cyclone in Bangladesh. *Focus on Gender 1*(1), 34–39.

Bent, D. (1995). Minimising business interruption: The case for business continuance planning. In A. G. Hull & R. Coory (Eds.), *Proceedings of the natural hazards management workshop 1995.* Institute of Geological and Nuclear Sciences.

Bernstein, P. L. (1996). *Against the gods: The remarkable story of risk.* John Wiley & Sons.

Bernkopf, R. L., Brookshire, D. S., & Thayer, M. A. (1990). Earthquake and volcano hazard notices: An economic evaluation of changes in risk perception. *Journal of Environmental Economics and Management, 18,* 35–49.

Blaikie, P., Cannon, T., Davis, I., & Wisner, B. (1994). *At risk–natural hazards: People's vulnerability and disasters.* Routledge.

Bolin, R. (1988). Natural disasters. In R. Gist & B. Lubin (Eds.), *Psychological aspects of disasters* (pp. 61–86). Wiley.

Britton, N. R. (1992a). Developing an understanding of disasters. *ANZ Journal of Sociology, 22,* 254–271.

Britton, N. R. (1992b). Uncommon hazards and orthodox emergency management. *International Journal of Mass Emergencies and Disasters, 10,* 329–348.

Britton, N. R. (1993). Seismic risk, utility lifeline vulnerability, and hazard mitigation in Wellington. In N. R. Britton & J. Oliver (Eds.), *Catastrophe insurance for tomorrow: Planning future adversities.* Sterling Offices Ltd.

Britton, N. R. (1995). Development of the emergency social system in Kobe following the earthquake. Paper presented to the Wellington Earthquake Lifelines Group. Wellington, New Zealand.

Britton, N. R., & Lindsay, J. (1995). Integrating city planning and emergency preparedness. *International Journal of Mass Emergencies and Disasters, 13,* 93–106.

Britton, N. R., & Oliver, J. (1991). The meaning of hazard. In N. R. Britton & J. Oliver (Eds.), *National and technological hazards: Implications for the insurance industry.* Sterling Offices Ltd.

Britton, N. R., & Oliver, J. (1993). Changing societal hazardousness and increasing social vulnerability: Whose responsibility? In N. R. Britton & J. Oliver (Eds.), *Catastrophe insurance for tomorrow: Planning for future adversities.* Sterling Offices Ltd.

Britton, N. R. (1999). Whither the emergency manager? *International Journal of Mass Emergencies and Disasters, 17,* 223–235.

Buckle, P. (1995). Community-based management of social disruption following disasters. *Australian Journal of Emergency Management, 10,* 31–38.

Burby, R. J. (Ed.). (1998). *Cooperating with nature: Confronting natural hazards with land-use planning for sustainable communities.* National Academy Press.

Burger, J. M., & Palmer, M. I. (1992). Changes in and generalisation of unrealistic optimism following experiences with stressful events: Reactions to the 1989 California earthquake. *Journal of Personality and Social Psychology, 18,* 39–43.

Burton, I., Kates, R. W., & White G. (1993). *The environment as hazard* (2nd ed.). Guilford Press.

Cook, J. D., & Bickman, L. (1990). Social support and psychological symptomatology following a natural disaster. *Journal of Traumatic Distress, 3,* 541–556.

Cohen, R. E., & Ahearn, F. L., Jr. (1980). *Handbook for mental health care of disaster victims.* The John Hopkins University Press.

Coile, R. C. (1997). The role of amateur radio in providing emergency electronic communication for disaster management. *Disaster Prevention and Management, 6,* 176–185.

Collins, J. M., & Ruelfi, T. M. (1996). *Strategic risk: A state-defined approach.* Kluwer Academic Publisher.

Cosgrove, J. (1996). Decision making in emergencies. *Disaster Prevention and Management, 5,* 28–35.

Covello, V. T., & Merkhofer, M. W. (1993). *Risk assessment methods: Approaches for assessing health and environmental risks.* Plenum Press.

Creasy, C. (1994). Emergency communications. *Fire Engineering, 147,* 40–42.

Cronin, S. (2020). *130.705 Emergency management, study guide, part 1.* Institute of Natural Resources, Massey University: Palmerston North, New Zealand.

Cuthbertson, B. H., & Nigg, J. M. (1987). Technological disaster and the nontherapeutic community. *Environment and Behavior, 19,* 462–483.

Cutter, S. (1995). The forgotten casualties: Women, children, and environmental change. *Global Environmental Change, 5*(3), 181–194.

Davies, J. C. (Ed.) (1996). *Comparing environmental risks: Tools for setting government priorities.* Resources for the Future.

Drabek, T. E. (1986). *Human system responses to disasters.* Springer Verlag.

Drabek, T. E. (1987). *The professional emergency manager: Structures and strategies for success.* [Monograph 44] Natural Hazards Research and Applications Information Center.

Drabek, T. E. (1990). *Emergency management: Strategies for maintaining organizational integrity.* Springer Verlag.

Drabek, T. E., & Hoetmer, G .J. (1991). *Emergency management: Principles and practices for local government, municipal management series.* International City Management Association.

De Ville De Goyet, C. (1993). Post-disaster relief: The supply–management challenge. *Disasters, 17,* 169–176.

Dudasik, S. W. (1980). Victimisation in natural disaster. *Disasters, 4*(3), 329–338.

Dynes, R. R. (1994). Community emergency planning: False assumptions and inappropriate analogies. *International Journal of Mass Emergencies and Disasters, 12,* 141–158.

Emergency Action Plans and Fire Protection Plans: OSHA Regulation 29 CFR 1910.38. (1999). The Hartford, Loss Control Department, Technical Information Paper Series: Hartford, Connecticut.

Eranen, L. & Liebkind, K. (1993). Coping with disaster: The helping behaviour of communities and individuals. In J. P. Wilson & B. Raphael (Eds.), *International handbook of traumatic stress syndromes.* Plenum Press.

Farley, J. E., Barlow, H. D., Finkelstein, M. S., & Riley, L. (1993). Earthquake hysteria before and after: A survey and follow-up on public response to the Browning forecast. *International Journal of Mass Emergencies and Disasters, 11,* 305–321.

Fothergill, A. (1996). Gender, risk and disasters. *International Journal of Mass Emergencies and Disasters, 14,* 33–56.

Gardiner, W. (1995, March 27–29). *Earthquakes: A maori dimension.* Paper presented at the Proceedings of the Challenge of Rebuilding Cities Conference, Wellington, New Zealand.

Gist, R., & Lubin, B. (1999). *Response to disaster.* Taylor and Francis.

Godschalk, D. R. T., Beatley, T., Berke, P., Brower, D. J., & Kaiser, E. J. (1999). *Natural hazard mitigation: Recasting disaster policy and planning.* Island Press.

Green, B. L. (1993). Identifying survivors at risk: Trauma and stressors across events. In J. P. Wilson & B. Raphael (Eds.), *International handbook of traumatic stress.* Plenum Press.

Green, B. L., & Solomon, S. D. (1995). The mental health impact of natural and technological disasters. In J. R. Freedy & S. E. Hobfall (Eds.), *Traumatic stress: From theory to practice.* Plenum Press.

Green, B. L., Gleser, G. C., Lindy, J., Grace, M., & Leonard, A. (1995). Age-related reactions to the Buffalo Creek Dam collapse: Second decade effects. In P. E. Rushkin & J. A. Talbott (Eds.), *Aging and post-traumatic stress disorder.* American Psychiatric Press.

Green, C. H., Tunstall, S. M., & Fordham, M. H. (1992). The risk from flooding: Which risks and whose perceptions? *Disasters, 15,* 227–236.

Gregory, G. (1995). Persuading the public to make better use of natural hazard information. *Prometheus, 13,* 61–71.

Hayes, B. D. (2004). Interdisciplinary planning of non-structural flood hazard mitigation. *Journal of Water Resources Planning and Management, 130,* 15–25.

Heath, R. (1995). The Kobe earthquake: Some realities of strategic management of crises and disasters. *Disaster Prevention and Management, 4,* 11–24.

Helm, P. (1996). Integrated risk management for natural and technological disasters. *Tephra, 15,* 5–13.

Hewitt, K. (1995). Excluded perspectives in the social construction of disaster. *International Journal of Mass Emergencies and Disasters, 13,* 317–339.

Hewitt, K. (1997). *Regions of risk: A geographical introduction to disasters.* Longman.

Hood, C., & Jones, D. K. C. (1996). *Accident and design: Contemporary debates in risk management.* UCL Press.

Institution of Civil Engineers. (1995). *Megacities: Reducing vulnerability to natural disasters.* Thomas Telford.

Johnston, D. M., & Houghton, B. F. (1995). Secondary school children's perceptions of natural hazards in the central North Island. *New Zealand Journal of Geography, 99,* 18–26.

Johnston, D. M., Bebbington, M., Lai, C.-D., Houghton, B. F., & Paton, D. (1999). Volcanic hazard perceptions: Comparative shifts in knowledge and risk. *Disaster Prevention and Management, 8,* 118–126.

Kasperson, R. E., & Pijawka, K. D. (1985, January). Societal responses to hazards and major hazard events: Comparing natural and technological hazards. *Public Administration Revue, 45* (Special Issue), 7–19.

Kasperson, R. E., Renn, O., Slovic, P., Brown, H. S., Emel, J., Goble, R., Kasperson, J. X., & Ratick, S. (1988). The social amplification of risk. *Risk Analysis, 8,* 177–187.

Kunreuther, H., & Gowda, R. (Eds.). (1990). *Integrating insurance and risk management for hazardous wastes.* Kluwer Academic Publishers.

LaPlante, J. M. (1988). Recovery following disasters: Policy issues and dimensions. In L. K. Comfort (Ed.), *Managing disasters: Strategies and policy perspectives* (pp. 217–235). Duke University Press.

Lange, S. (1998). Disaster planning: The challenge within. *Risk Management, 45,* 34–37.

Lebedun, M., & Wilson, K. K. (1989). Planning and integration disaster response. In R. Gist & B. Lubin (Eds.), *Psychological aspects of disasters* (pp. 268–279). New York.

Lehman, D. R., & Taylor, S. E. (1987). Date with an earthquake: Coping with a probable, unpredictable disaster. *Personality and Social Psychology Bulletin, 13,* 546–555.

Lindell, M. K. (1994). Perceived characteristics of environmental hazards. *International Journal of Mass Emergencies and Disasters, 12,* 303–326.

Lindell, M. K., & Perry, R. W. (1992). *Behavioural foundation of community, emergency planning.* Taylor & Francis.

May, P. J. (1988). Disasters recovery and reconstruction. In L. K. Comfort (Ed.), *Managing disaster: Strategies and policy perspectives* (pp. 236–254). Duke University Press: Durham.

May, P. J. (1997). Addressing natural hazards: Challenges and lessons for public policy. *Australian Journal of Emergency Management, 11,* 30–37.

Quarantelli, E. L. (1986). Research findings on organisation behaviour in disasters and their applicability in developing countries (Preliminary Paper #107). Disaster Research Center, University Delaware.

3

Disaster Response

1.0 INTRODUCTION 88
2.0 FEDERAL AND REGIONAL RESPONSE SUPPORT 91
2.1 Department of Homeland Security National Response Framework 91
2.2 National Incident Management System 94
2.3 Incident Command System 94
2.4 National Preparedness Guidelines 96
2.5 Federal Response Support—U.S. Environmental Protection Agency 96
2.6 Federal Response Support—Other Agencies 97
2.6.1 Federal Bureau of Investigation 97
2.6.2 Department of Homeland Security 98
2.6.3 Centers for Disease Control and Prevention 98
2.7 Regional Support 99
2.7.1 Water/Wastewater Agency Response Network 99
2.7.2 Risk Management Planning 100
2.7.3 Incident Command Team 101
2.8 Communication 101
2.8.1 Communication Training 102
2.8.2 Contact Lists and Key Documents 103
2.8.3 Maintaining Credibility 104
3.0 UTILITIES—RESPONSE 105
3.1 Initial Response Actions 105
3.2 Planning as Incident Unfolds 106
3.3 Initial Response 106
3.3.1 The Start of Each Planning Cycle 116
3.3.2 Preparing for and Conducting the Tactics Meeting 118
3.3.3 Preparing for the Planning Meeting 119
3.3.4 Conducting the Planning Meeting 119
3.3.5 Incident Action Plan Preparation a nd Approval 119
3.3.6 Operations Period Briefing 120
3.3.7 Execute Plan and Assess Progress 120
3.4 Incident-Specific Procedures 120
4.0 INCORPORATING SAFETY TO DISASTER RESPONSE 124
5.0 COMMUNICATION: PLANNING BECOMES DOING 125

5.1 Say What Can Be Said 126
5.2 Create a Holding Statement 126
5.3 Tell More of What Can Be Said 128
5.4 Prepare for Two-Way Conversations; Social Media Can Be a Two-Edged Sword 128
5.5 Express Concern for Victims 128
5.6 Respond to Questions the Media Will Ask 129
5.7 Avoid Saying "No Comment" 129
5.8 Identify Affected Parties 129
5.9 Prepare Written Statements 130
5.10 Conduct a Media Briefing 131
5.11 Take Photographs 131
5.12 Additional Tips 132
6.0 RESOURCES 133
7.0 REFERENCES 153

1.0 INTRODUCTION

Whenever the unthinkable happens, all water and wastewater utilities must be prepared to respond, but nobody has to feel alone. There is more support for utilities than ever now. The federal government has committed significant resources to national emergency response and preparedness and states have additional supports. Numerous industry guidelines are also available, which cover vulnerability assessments, risk analyses, preparedness, mitigation, training, response, and recovery activities, among other steps in the cycle.

Any event happens as part of an entire preparedness cycle that should have begun with planning and continue through corrective and preventive action. Incidents go through a life cycle beginning with notification, assessment, and emergency response activation, and ending with stand-down and return to service. Figure 3.1 is an example of an incident life cycle.

Table 3.1 outlines the four phases of emergency response used by the Federal Emergency Management Agency (FEMA).

Emergency conditions can result from natural disasters, human error, equipment and structural failures, chemical releases, fire or explosion, probable permit violations, terrorism, or sabotage. The number one goal of a utility manager is to provide timely and effective response to any emergency condition and to protect the lives and health of facility personnel and the general public, protect the environment, maintain service operation, and ensure timely notification of appropriate internal and external officials.

The chapter is divided into sections that will cover a journey from supports that are available at the national level and the local level in response to an event. This chapter does not detail incident-specific emergency response procedures. As an important note, each utility must develop its own

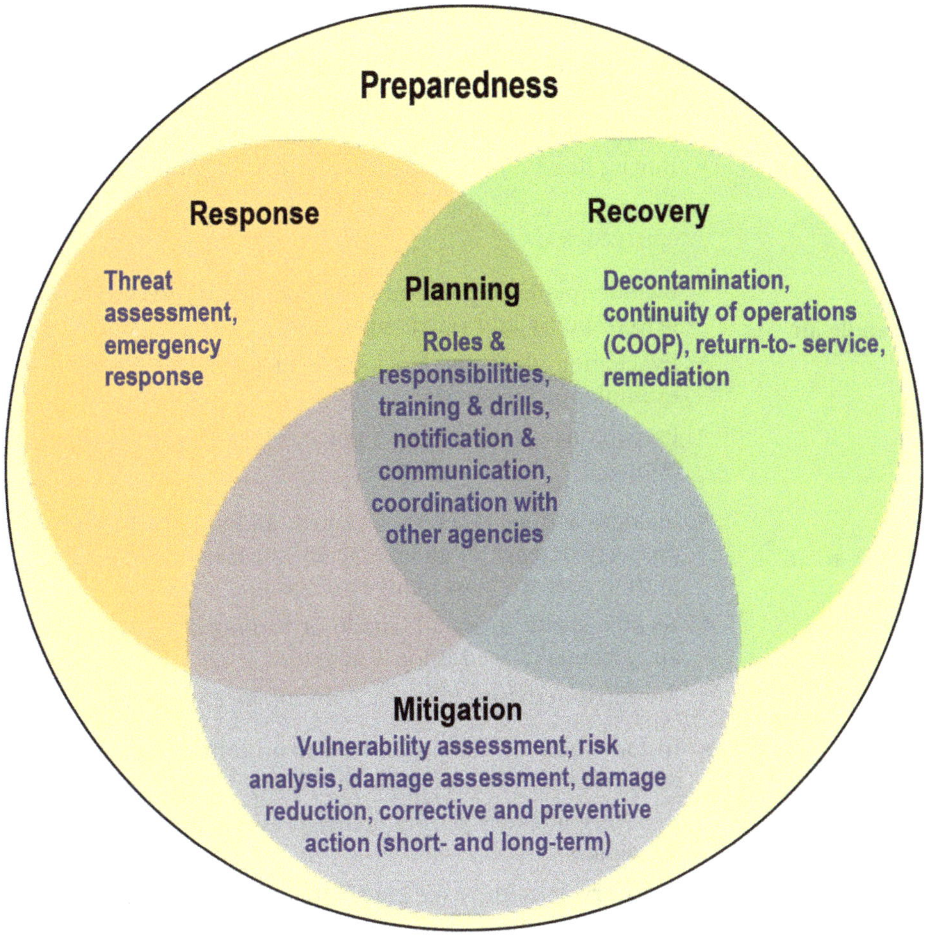

FIGURE 3.1 Incident Life Cycle

incident-specific emergency response procedures, which should be developed by individual organizations based on their risk analyses and circumstances, using some of the many industry guidelines. The Federal Emergency Management Agency's National Incident Management System (http://www.fema.gov/national-incident-management-system) Web site is replete with guidance and offers online training and information on certifications. It is worth the time to become familiar with this entire website.

The terms *emergency*, *incident*, and *event* are used interchangeably in this section, to mean "any condition that could or has resulted in serious personnel injury or death, results in an environmental or health threat (probable legal or regulatory violation) to the general public, or could cause substantial damage to the organization's infrastructure or operations."

TABLE 3.1 Four Phases of Emergency Management

Mitigation Preventing future emergencies or minimizing their effects	• Includes any activities that prevent an emergency, reduce the chance of an emergency happening, or reduce the damaging effects of unavoidable emergencies. • Buying flood and fire insurance is a mitigation activity. • Mitigation activities take place before and after emergencies.
Preparedness Preparing to handle an emergency	• Includes plans or preparations made to save lives and to help response and rescue operations. • Evacuation plans and stocking food and water are examples of preparedness. • Preparedness activities take place before an emergency occurs.
Response Responding safely to an emergency	• Includes actions taken to save lives and prevent further property damage in an emergency situation. Response is putting preparedness plans into action. • Seeking shelter from a tornado or turning off gas valves in an earthquake are response activities. • Response activities take place during an emergency.
Recovery Recovering from an emergency	• Includes actions taken to return to a normal or an even safer situation following an emergency. • Recovery includes getting financial assistance to help pay for repairs. • Recovery activities take place after an emergency.

Most of an emergency management plan (EMP) is not referenced during an actual event. The plan is the basis for training and drills, and to capture lessons learned through continuous updates that cycle back into training. The items most critical to effective response are personnel training based on realistic standardized protocols, professional experience, and checklists to ensure nothing critical is missed during an event. Field or quick-reference guides are useful for the latter.

Effective response relies on preparedness and ensuring those standard protocols are complete, up to date, and followed, including the following:

- Everyone knowing their roles and responsibilities during an emergency, which often are different from the standard organizational structures. Egos need to move aside.

- Using what was learned during training and drills, and following procedures. When those fall short, relying on professional experience and knowing who has it.
- Following established notification and communication procedures so nobody is left out of the loop and messages are not crossed.
- Respecting a pre-established decision-making process throughout the event. This includes a common understanding of threat level designations for accurate initial and ongoing threat assessments and ensures proper response to escalating or de-escalating events.
- Continuously monitoring the threat levels and engaging outside agencies and organizations as demanded.
- Including all appropriate levels of action in the response, such as immediate on-site organizational response to local, state, and federal response agencies.

The following sections focus first on the federal supports because these structures can form a blueprint for establishing an organizational response plan. State and local supports often cascade down from federal programs or otherwise mirror those programs. For example, most states have departments of homeland security (DHSs) and emergency management agencies (EMAs). Counties and municipalities often have their counterparts. Several federal programs have regional locations, such as fusion centers. Each state has its own version of emergency management programming with response capabilities and procedures, but the vigorous trend among states, local jurisdictions, and individual agencies is to synchronize these efforts and follow national DHS/FEMA standards and protocols, as explained below.

The last sections focus on typical timelines and actions preparing for and during an emergency at a water/wastewater utility, including initial response, communication, and planning as an emergency unfolds.

2.0 FEDERAL AND REGIONAL RESPONSE SUPPORT

2.1 Department of Homeland Security National Response Framework

The National Response Framework (NRF) (http://www.fema.gov/national-response-framework), published in January 2018 by the U.S. Department of Homeland Security, is a set of guiding principles that enable all response partners and stakeholders to prepare for and provide a unified national response to hazards of all kinds. It was developed with input from partner stakeholders: federal agencies, states, local response organizations, tribes, and the private sector, all of whose roles are addressed.

The flow of federal guidance is NRF → National Incident Management System (NIMS) → Incident Command System (ICS).

Based on best practices identified following Hurricane Katrina in 2005, the NRF describes specific authorities and relationships for managing incidents that range from the serious, but strictly local, to large-scale terrorist attacks or catastrophic natural disasters. It provides guidance on how roles align for efficiency and effectiveness during response operations. It is not a response plan, but a basis for developing response plans at all levels of government. A critical principle of the NRF is that all incidents are managed locally.

The NRF presents the key response principles, identifies the participants and their roles and responsibilities, and describes structures that guide the nation's response operations. It provides all response personnel, including water/wastewater sector officials, with guidance and information relating to the following:

- Those who are involved with emergency management activities at each level of government.
- What the nation collectively does to respond to incidents; for example, state departments of transportation will provide resource support to local governments responding to incidents. When the local government's resources are exhausted, the state will provide additional resources, and then the federal government will provide support to the state as needed.
- How, at all levels, the nation effectively manages an all-hazards response; for example, by establishing a joint field office with state and federal personnel in direct support to local responders.

The NRF also lists expected capabilities and membership of 15 emergency support functions, including public works—water, stormwater, sewer, and wastewater services:

1. Transportation;
2. Communications;
3. Public works and engineering;
4. Firefighting;
5. Emergency management;
6. Mass care, emergency assistance, housing, and human services;

7. Logistics management and resource support;
8. Public health and medical devices;
9. Search and rescue;
10. Oil and hazardous materials response;
11. Agriculture and natural resources;
12. Energy;
13. Public safety and security;
14. Long-term community recovery; and
15. External affairs.

Figure 3.2 from the NRF demonstrates the support flow when mutual aid or state and federal assistance are sought.

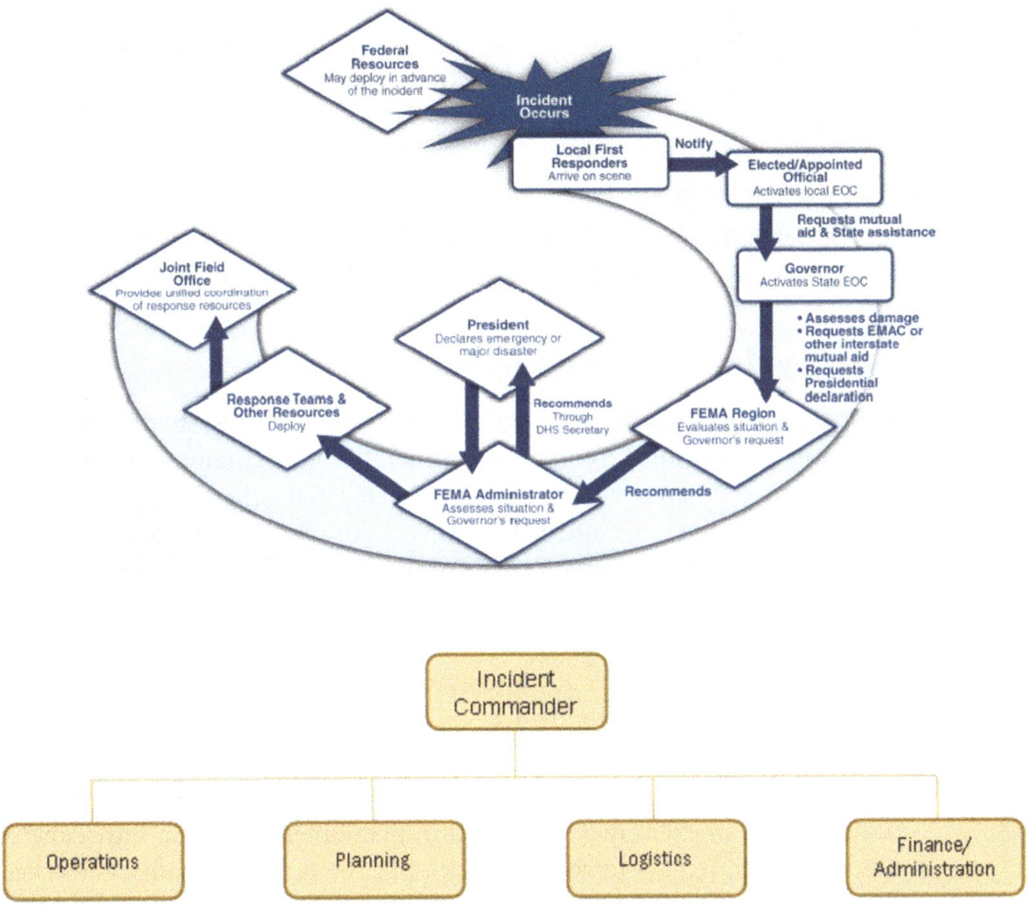

FIGURE 3.2 National Response Framework (NRF) Support

2.2 National Incident Management System

The Federal Emergency Management Agency's NIMS (http://www.fema.gov/national-incident-management-system) is a companion document to the NRF and goes into practical detail to help organizations implement the NRF locally. It provides a systematic, proactive approach for use by departments and agencies at all levels of government, nongovernmental organizations, and the private sector. The approach allows them work seamlessly to prevent, protect against, respond to, recover from, and mitigate (lessen) the effects of incidents. The NIMS works hand in hand with the NRF to reduce the loss of life, property damage, and harm to the environment.

It provides the template for incident management; the NRF provides the structure and mechanisms for national-level policy for incident management.

The NIMS is intended to accomplish the following:

- Be used in response to all types of incidents, hazards, and emergencies regardless of cause, size, location, or complexity;
- Improve coordination and cooperation between public and private entities in a variety of incident management activities; and
- Provide a common organizational structure and language for incident management.

In other words, the NIMS is intended ensure all organizations react similarly when responding to all types of emergencies.

2.3 Incident Command System

Though ICS (http://www.fema.gov/incident-command-system) is a federal program, it is a critical tool for local officials and is gaining popularity as a best practice. When it was developed in the 1970s, there was no standard system to manage emergency situations. Many Americans can recall cases in which responders were at cross-purposes or could not communicate on a common bandwidth.

An ICS allows organizations to establish standardized systems understood by everyone—internally and externally. Among other benefits, it provides multiple levels of readily accessible training for every employee, much of it online.

Under an ICS, incidents are addressed as planned events or as unforeseen occurrences. An event may be a natural phenomenon (e.g., a hurricane) or human-caused (e.g., a terrorist attack), but the defining quality is the need for emergency response or action to prevent or minimize adverse effects on human health and damage to property and/or natural resources.

The system was developed around the following principles:

- Flexibility. ICS has a flexible organizational structure, allowing the size of the response to expand or contract to meet the requirements of an incident;
- Standardization. The standard format and terminology of an ICS allows personnel from different agencies and jurisdictions to converge into a single management structure; and
- Cost-effectiveness. Because costs associated with emergency responses can be quite high, an ICS seeks to be cost-effective in its structure.

Since its inception, ICS has been used to manage emergency situations from small hazardous materials releases, to natural disaster responses, to wide-area search and rescue missions. It has been adopted or endorsed by numerous agencies, including FEMA, the U.S. Environmental Protection Agency (U.S. EPA), and the Occupational Safety and Health Administration (OSHA). Furthermore, it has been adopted by state EMAs and it is regularly used by emergency responders at water/wastewater utilities.

Under ICS, roles and positions in an incident response are divided into five functional areas, briefly defined as follows (see Figure 3.2):

- Incident Command: Responsible for overall incident response. Sets priorities and objectives. Oversees information dissemination, contact with external agencies, and health/safety aspects of the response.
- Operations: Responsible for developing tactical objectives and overseeing the operations organization and resource allocation. Implements tactical actions to achieve plan objectives.
- Planning: Responsible for developing the action plan to accomplish the objectives. Collects and evaluates incident information.
- Logistics: Responsible for providing support, resources, and services to meet the needs of the incident response.
- Finance/Administration: Responsible for tracking incident costs. Provides accounting and procurement services, records employees' time, and performs cost analyses.

The organization of ICS is flexible, allowing the size of the response team to expand and contract as response needs change. Each response involves, at a minimum, an incident commander. In a small incident, the incident commander might carry out all of the functions described above. Or the incident commander may carry out some of the functions, but assign personnel to accept responsibility for performing other functions. Thus, the size of the response team grows as the incident commander determines that managing the incident requires more resources.

2.4 National Preparedness Guidelines

Other supports provided by U.S. DHS are the National Preparedness Guidelines (http://www.dhs.gov/national-preparedness-guidelines) and various companion documents, including the Target Capabilities List (http://www.fema.gov/pdf/government/training/tcl.pdf) and the Universal Task List (https://www.rkb.us/contentdetail.cfm?content_id=185590).

The following lists the purposes of the guidelines:

- Organize and synchronize national (including federal, state, local, tribal, and territorial) efforts to strengthen national preparedness;
- Guide national investments in national preparedness;
- Incorporate lessons learned from past disasters into national preparedness priorities;
- Facilitate a capability-based and risk-based investment planning process; and
- Establish readiness metrics to measure progress and a system for assessing the nation's overall preparedness capability to respond to significant events, especially those involving acts of terrorism.

The Universal Task List is a menu (in Excel) of approximately 1600 unique tasks and a common vocabulary for use with national planning scenarios, which can be used by water/wastewater agencies to develop response plans.

2.5 Federal Response Support—U.S. Environmental Protection Agency

U.S. EPA has invested enormous resources in developing guidelines and practical tools for the water and wastewater sector. Its website on emergency/incident planning, response, and recovery (http://water.epa.gov/infrastructure/watersecurity/emerplan/index.cfm) is replete with training, exercise, and preparedness planning materials, and recovery tools.

Figure 3.3 demonstrates U.S. EPA's approach to a response, which mirrors FEMA's. Many organizations are adding "Awareness" as a first step.

Among U.S. EPA notable resources are the following:

- The Water Security Publications Database (http://cfpub.epa.gov/safewater/watersecurity/publications.cfm). Among other publications, it includes case studies, information on legislation, outreach materials, webcasts, and NIMS training for the water sector.

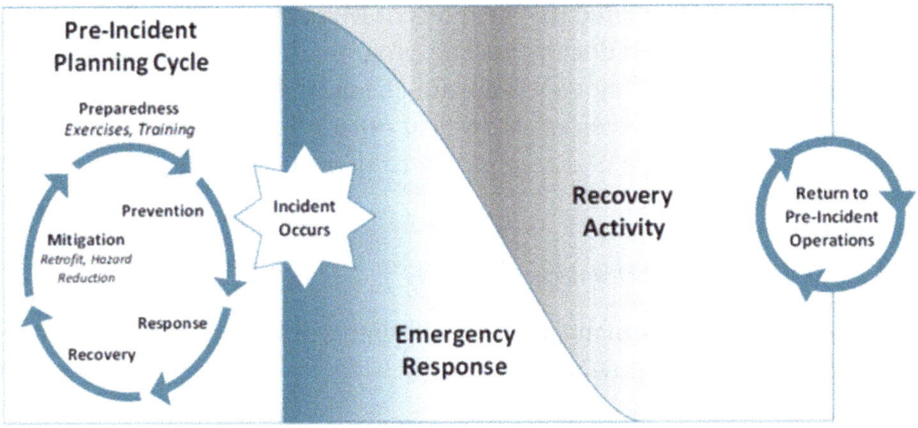

FIGURE 3.3 U.S. EPA Approach to a Response

- Vulnerability Self-Assessment Tool (VSAT) (http://water.epa.gov/infrastructure/watersecurity/techtools/vsat.cfm). This is a risk assessment software tool for water, wastewater, and combined utilities of all sizes. The tool helps this sector perform security threat and natural hazards risk assessments, and updates utility emergency response plans. U.S. EPA, in collaboration with water sector partners (including the National Association of Clean Water Agencies) and stakeholders, updated VSAT to be consistent with the risk analysis and management framework for critical asset protection. The tool includes a user interface, an enhanced natural disaster threat assessment process, and a revised risk assessment approach.
- Climate Resilience Evaluation and Awareness Tool (http://water.epa.gov/infrastructure/watersecurity/climate/creat.cfm). This software tool helps water sector owners and operators understand potential climate change threats and assess related risks. Climate change was linked to increasingly extreme, frequent, and unpredictable weather events. Although this tool is more in line with long-term planning, putting it to use can mitigate emergencies and reduce the need for extreme response efforts.

2.6 Federal Response Support—Other Agencies

2.6.1 Federal Bureau of Investigation

The Federal Bureau of Investigation (FBI) has a robust program to address terrorism, including weapons of mass destruction (WMD) and criminal use of chemicals (http://www.fbi.gov/about-us/investigate/terrorism/wmd). Although water/wastewater utilities have taken steps since 2001 to reduce vulnerability from chemicals and have implemented security measures, they remain at risk.

In 2006, the FBI created the Weapons of Mass Destruction Directorate. This website and related documents and programs include guidance on preparing for and responding to WMD and chemical attacks. There is a local WMD coordinator for each of its 56 field offices that will work with local authorities on reporting and responding to threats. The FBI also conducts workshops on water terrorism.

2.6.2 Department of Homeland Security

Noteworthy programs managed by DHS that would interest water sector officials include the following:

- Fusion Centers (http://www.dhs.gov/state-and-major-urban-area-fusion-centers). These are regional focal points to centralize threat-related information for planning for and during real-time events. They receive, analyze, gather, and share information among all levels of government and private sector partners and stakeholders.
- Protective Security Advisors (PSAs) (http://www.dhs.gov/protective-security-advisors). The PSAs are trained experts in critical infrastructure protection and vulnerability mitigation. They conduct specialized site visits and can work with utilities to provide guidance on protecting critical infrastructures.
- Hazardous Response Program (http://www.dhs.gov/hazardous-response-program). This Federal Protective Service program provides response to credible chemical, biological, radiological, nuclear, and high-yield explosive (CBRNE) threats or incidents. It conducts initial investigations of suspicious or threatening CBRNE incidents and complies with OSHA and National Fire Protection Association guidance and regulations. Response teams are available to utilities.

2.6.3 Centers for Disease Control and Prevention

The Centers for Disease Control and Prevention (CDC) have many recommendations (http://emergency.cdc.gov/) that can be translated into water/wastewater utility-specific incident response procedures. Several information-sharing tools are available at the federal level:

- The Water Information Sharing and Analysis Center (https://portal.waterisac.org/web/) was authorized by Congress in 2002 to keep drinking water and wastewater utility mangers informed of potential risks to the nation's water infrastructure from contamination, terrorism, and cyberthreats. It is created and managed by the water sector and links members through a secure online portal.

- Homeland Security Information Network (http://www.dhs.gov/homeland-security-information-network) brings together emergency services professionals in "communities of interest" and provides situational awareness tools.
- The Community Emergency Response Teams Program (https://www.citizencorps.gov/cert/) educates people about disaster preparedness for hazards that may affect their area and helps neighborhoods and localities work and respond together.

2.7 Regional Support

Emergency support in most jurisdictions goes two ways: Support from the agency to other agencies and organizations, and support from them to their local governments for incidents on their facilities.

In some cases, utilities will play a supporting role for incidents that affect other agencies. Most governments have response plans that cover these scenarios.

In other cases, such as onsite fires or floods, utilities may need the support of other agencies. Depending on the incident, a joint information system may need to be established to coordinate information dissemination. For incidents that involve multiple agencies, a unified command may need to be established.

Planning requires that relationships be established in advance with these organizations so emergency response is efficient and effective. Joint training should be conducted and focus on mutually developed priorities based on risk analyses conducted by each organization. Related lines of communication, areas of responsibility, and associated procedures and resource banks should be established, understood, and tested before an emergency as part of the planning process. Post-incident reviews and corrective and preventive action should be done in collaboration with affected agencies/organizations as part of the continuous improvement process, feeding back into planning.

2.7.1 Water/Wastewater Agency Response Network

Regional utilities have voluntarily agreed to help other utilities prepare for the next natural or human-caused emergency through the regional Water/Wastewater Agency Response Network (WARN) (http://www.awwa.org/resources-tools/water-knowledge/emergency-preparedness/water-wastewater-agency-response-network.aspx). The network provides mutual aid assistance in the form of personnel, equipment, materials, and other services. For example, the National Capital Region (NCR) of the United States consists of Washington, D.C., and the surrounding counties and independent cities in Maryland and Virginia. The NCR WARN Web site is hosted by the Metropolitan Washington Council of Governments.

2.7.2 Risk Management Planning

Good EMPs are based on a risk management program that includes a vulnerability assessment, a risk analysis, threat identification, and a business effect analysis. Risk management terms are defined in Table 3.2.

Here is one way to approach the steps in identifying possible crises:

1. Identify several types of events that would be a crisis if they occurred. A crisis is a critical event—a turning point. If not handled properly, it could turn into a disaster for the organization and negatively affect its ability to accomplish its mission.
2. In considering crisis-precipitating events, think in terms of consequence and likelihood. An event of great consequence that is highly unlikely should not get as much managerial concern as one that is of great consequence and is highly likely.

TABLE 3.2 Risk Management Terms

Term	Description
Asset	What we are trying to protect. These are people, property, and information.
Threat	What we are trying to protect against. These are hazards to which assets are susceptible, caused by natural events, accidents, or actions intended to cause harm.
Vulnerability	A weakness or gap in our protection efforts. Threats (actual, conceptual, or inherent) may exist, but if there are no vulnerabilities, there is little or no risk. If there is vulnerability, but no threat, then there is little to no risk.
Risk	The adequacy of physical, environmental, security, or other controls to protect against a threat; a function of threats exploiting vulnerabilities to obtain, damage, or destroy assets. All facilities face certain levels of risk associated with assorted threats. Risk assessment includes criteria ratings based on history, vulnerability, threat, and probability.
	Asset + Threat + Vulnerability = Risk

3. Look at disasters that have befallen similar organizations. Think not only about the events and what was done to recover, but also about what the organization's spokesperson should say to the media or an affected stakeholder in future occurrences.
4. Talk with the people in the organization about their thoughts about the organization's most serious vulnerabilities. How likely are they? What would the consequences be if one happened?

2.7.3 Incident Command Team

It is important to establish a command structure that identifies the key individuals and clearly defines their roles and responsibilities to avoid confusion and ensure appropriate training so they may effectively manage an emergency situation. It is equally important to establish a structure that can be easily adapted if the incident or event evolves into a multiagency or multijurisdictional response, in which utility personnel become agency representatives on a larger response team or organization.

The incident command system that NIMS encourages is composed of managers and senior staff in five functional areas who respond to various emergency or potentially dangerous situations, such as chemical leaks and spills, fires, floods, and terrorist activities. These functional areas are overall command, logistics, operations, finance and administration, and planning. Their leaders make up the incident command team, with the addition of a safety and security officer, public information officer (PIO), legal officer, and liaison officer, headed by the incident commander (or similar crisis manager).

Although a single incident commander typically handles the command function, ICS organization may be expanded into a unified command.

2.8 Communication

Does the organization have a crisis communication plan? If it has one, is it up to date? Following are items that should be contained in the crisis communication plan:

- Identify the PIO or crisis spokesperson, his or her backup, that person's backup, and everyone's roles and responsibilities in a crisis. The communication plan should outline to whom each person reports and who has authority to clear information for release to the stakeholders. (Hint: The incident control system has PIO reporting to the incident commander.)
- State and disseminate the media policy. A good basis is maximum disclosure = minimum delay. The policy should state that the designated

spokesperson is the only person who speaks for the organization to the media. Speaking with one voice in a crisis is critical to the organization and to those who are receiving the message. There must be no confusion about what the organization is saying, who is saying it, and the authority with which the spokesperson speaks, especially if people's lives or property might be at stake.

- Create a communication flowchart. The PIO's job in a crisis is to get the right information to the right people at the right time so the right decisions are made. Organizationally, the PIO must report directly to the manager in charge of resolving the crisis. Intermediaries or go-betweens between the crisis manager and the PIO are not allowed because complete clarity, trust, and understanding between them are essential.
- Differentiate communication from notification. Notification involves immediate alerts to incident responders or others directly responsible for managing the incident and its effects. Incident communication is conducted throughout the life of the incident and is aimed at ensuring all incident personnel, the public, and other stakeholders are aware of what is happening so they can respond or take preventive measures, or report information accurately.
- Ensure all communication personnel are trained. See section below.
- Maintain updated contact lists. See section below.
- Establish media staging. Determine where media representatives will gather and where their satellite trucks will be allowed to park. A boardroom or similar public space with restrooms and access to telephones and electrical outlets is ideal. The media holding room should not be next to (or near) the command-post room, the PIO's office, or other staff offices. Media representatives cannot be allowed to wander freely or interfere with working staff as they resolve the crisis.

2.8.1 Communication Training

Train the PIO/spokesperson (and backups) ahead of time; it is too late to identify or train them after a crisis occurs. It is not unusual that they can get stuck off site during an emergency, especially during weather events that render roads impassable and impair telephone lines. Also, people go on vacations, get sick, go to conferences, quit, and are otherwise unavailable. The PIO does not have to serve as spokesperson. Smaller utilities often do not have a full-time PIO but they can designate and train someone (and backups) to take on this role before a crisis strikes.

Choose a potential spokesperson carefully. Some people do not speak clearly or are simply too frightened of public speaking to do well on camera. Pick as a spokesperson someone who is typically knowledgeable, speaks clearly, a quick study, and cool under pressure when the cameras and lights point his or her way.

Media training sessions should include a mock on-camera interview, which can help identify problems early and prepare the spokesperson or find an alternative. Train replacements when the spokesperson/PIO retires, leaves or changes jobs in the organization, or otherwise becomes unavailable.

The utility manager and technical experts who are most likely to have to go on camera should also be media-trained. These might include the operations manager, collection systems manager, lab manager, chief engineer, biosolids manager, or safety and security manager. Outside media trainers can often include three to five people in a class, which helps stretch the training budget.

2.8.2 Contact Lists and Key Documents

Does the organization have a key personnel list? This list should contain the names and telephone numbers (desk, cell, home, mountain getaway cabin) of all people on the key personnel list, including essential operations personnel for various kinds of threats. Keep the following documents updated:

- Key personnel contact list. This should include the incident commander and everybody on the incident management team, such as the PIO/spokesperson and the crisis manager, operations people, planners, logistics support, and those authorized by the organization to help resolve a problem. For example, in Colorado, a Colorado Discharge Permit System permit requires a chain-of-command list, which should be the starting place for a key personnel list.
- A one-page background document for the media that describes the organization, its mission, how it is constituted (i.e., as a department or division of a city or as a stand-alone special district), executives, population served, typical flows treated, and all of the facilities. This document should be written ahead of time and updated annually.

  If the crisis attracts a lot of media attention, it might be the first time some of the reporters covering the situation have ever paid any attention to the organization. They may need help just getting the basic facts straight (such as the organization's official name).
- An up-to-date media list. Even if the organization regularly sends news releases, it needs a list of media outlets (including telephone

numbers, fax numbers, and email addresses) of those that would reasonably be expected to cover the organization. Reporters, especially at smaller TELEVISION and radio stations and weekly newspapers and online outlets like the Patch, tend to change jobs frequently on their way to larger markets. Update the list at least annually.

- Research on previous crises. When preparing the media handout, research previous crises the organization has faced and what the outcomes were. Use this information to prepare the spokesperson and senior leaders. They need to know these events, their causes, and resolutions and whether the organization has a history of violations, especially in the safety or environmental arena. If a pattern similar to earlier events appears to be developing, expect to have to deal with questions about those events.
- A blank media query log sheet. When a crisis hits, questions will come rapidly and there will be no time to prepare this form. Include spaces for date and time the call was received, name of caller, media outlet, deadline, telephone number, fax number, email address, person taking call, person providing response, and question(s) asked. Make sure whoever answers telephones has blank log sheets on hand, fills them out, and passes them on promptly. This information will be valuable afterwards during post-event analysis.

2.8.3 Maintaining Credibility

Among people who have handled many crises, an old saying holds that crises come in three stages: (1) mess up, (2) 'fess up, and (3) dress up. Stage one—mess up—is whatever happened that precipitated the crisis. Stage two—'fess up—is the confession: Yes, we have had a wastewater leak (or whatever happened), we are sorry to have inconvenienced so many people, and here is what we are doing about it now. The third and final stage—dress up—is getting things working again and letting the public know that.

It is a mistake to believe that an event like a large wastewater surcharge in a residential neighborhood will not be noticed, yet organizations make this mistake every day. It is best to get out in front of the crisis. By being the first to report a potentially unpleasant situation, the organization's credibility is established as one that is honest, socially responsible, and on top of things.

What used to be the "news cycle" is now the "news moment." Reporters live in a world of constant, ongoing deadlines. Although there used to be a finite time—24 or 48 hours—between an event occurring and the media coverage and reactions, that time has now shrunk to virtually zero because of the pervasiveness of social media, such as Twitter, Facebook, and YouTube.

Getting out in front of a story allows the organization the opportunity to shape it, which does not mean "spin." It means quickly and truthfully providing all the information the organization has. The truth, however painful, must be told completely and quickly or the story will drag on forever as first one fact and then another is discovered. The media will peel back the layers, one at a time, and uncover new angles. Each new angle will generate a new story.

A *Bloomberg Businessweek* article (2007) said: "Messaging isn't 'spin.' Your key stakeholders have got to know they can count on you to tell the truth, no matter how unpleasant that truth is." The same article also said: "It's important to acknowledge people's anger or frustration, even when not accepting blame for wrongdoing. Avoid the temptation to minimize objections and complaints." This is especially true in the age of social media.

Also remember that traditional media see the world in terms of drama and conflict. Drama means the crisis is a story with a beginning, a middle, and an end. Conflict is more ambiguous. It can be man against man, man against nature, nature against man, or something else.

However it goes, there is a struggle involved, and the media will look for it. Do not be surprised when they do and the situation portrayed in a way that was not expected. Know also that in some newsrooms, there is a belief that someone must be responsible and there someone must be blamed.

3.0 UTILITIES—RESPONSE

In most emergencies, it rapidly becomes clear when the situation reaches crisis proportions. In some cases, it may be difficult to evaluate whether a crisis exists. The incident commander is responsible for determining threat levels, which in turn generate specific responses.

For purposes of most emergency plans, nonroutine events can be classified as emergencies if they meet one or more of the following conditions:

- Significantly interrupt normal operations;
- Endanger the public;
- Require an immediate, coordinated management response; and
- Have the potential for public interest and extensive media coverage.

3.1 Initial Response Actions

This section explains a typical response timeline to be followed by utility response personnel during an emergency event. Declaring a "state of emergency" and activating procedures in an EMP, including a formal incident response or incident response team, is a serious matter and requires an accurate assessment of the incident and potential for escalation.

Table 3.3 is an example of a threat level designation format that can be used to determine the appropriate level of response. Table 3.4 outlines a typical response time line.

3.2 Planning as Incident Unfolds

During any event, decision-making is relentless—choices must be made from the time a credible threat is identified, or when there is an initial response to an actual or impending event, through the entire cycle.

The "during-incident" planning process begins with identifying and assessing the threat level. It continues with implementation of formalized steps and staffing required to develop an incident action plan (IAP), which should be written using forms.

The primary phases of the planning process are essentially the same for the incident commander who develops the initial plan, for the incident commander and operations section chief revising the initial plan for extended operations, and for the incident management team developing a formal IAP.

During the initial stages of incident management, planners must develop a simple plan that can be communicated through concise verbal briefings. Frequently, this plan must be developed quickly and with incomplete situation information. As the incident management effort evolves over time, additional lead time, staff, information systems, and technologies enable more detailed planning and cataloging of events and "lessons learned."

Planning involves the following:

1. Evaluating the situation;
2. Developing incident objectives;
3. Selecting a strategy; and
4. Deciding which resources should be used to achieve the objectives in the safest, most efficient and cost-effective manner.

Table 3.5 lists steps related to planning during an incident. Figure 3.4 demonstrates the planning cycle that starts when a credible threat is identified, or when there is an initial response to an actual or impending event.

3.3 Initial Response

Planning begins with a thorough assessment that provides information that is needed to make initial management decisions.

TABLE 3.3 Response Matrix

Level	Definition	Types of effects	Response actions
1	**Near-normal operations** Localized events with little or no external assistance needed	• Minor or no injuries • Limited or no environmental effect • Minor damage to structures or infrastructure • < $100 000 in damages • Limited or no media scrutiny • Limited or no legal issues	• Utility open as usual • Notify Emergency Command Center (ECC)—ECC will make incident level 1 notifications— directly affected stakeholders only • Implement standard response protocols and procedures • No incident management team (IMT) activation • Notify local homeland security and emergency management agency (HSEMA)
2	**Localized emergency (low-level emergency)** Potential or actual localized emergency Limited external assistance needed.	• Moderate injuries potentially involving hospitalization • Some localized environmental effects • Localized damage to structures or infrastructure • Accidental spill or wastewater discharge violation with minimal localized effect • $100 000–$1M in damaged infrastructure • Attention from media and/or heightened concern from local authorities • Serious breach of regulations with mandatory investigations and reports to authorities	• Utility open as usual • Notify ECC—ECC will make incident level 2 notifications • Notify and identify the on-scene commander and/or the incident commander • Partial IMT activation considered by incident commander • Notify local HSEMA and consider notifications of affected stakeholders and critical infrastructure personnel • Consider establishing an incident command post • Consider requesting safety officer support • Consider requesting security officer support

(continued)

TABLE 3.3 Response Matrix (*Continued*)

Level	Definition	Types of effects	Response actions
3	**Beyond-localized emergency (high-level emergency)** Potential or actual emergency beyond a localized area External assistance needed.	• Severe injuries of one or more persons • Serious long-term environmental effects • Severe damage to critical infrastructure • Accidental spill or wastewater discharge violation with significant localized effect and/or potential effect on facility or river • $1M–$100M in damaged infrastructure • Significant adverse national media and regional authority attention • Significant fines and multiple litigations	• Utility facilities are open with essential personnel activated • Notify ECC—ECC will make incident level 3 notifications • Full IMT activation by incident commander • Notify and identify the on-scene commander and/or incident commander—incident commander to define the role of the IMT • Notify local HSEMA and make mandatory notifications to affected stakeholders and critical infrastructure personnel • Establish an incident command post • Conduct initial briefing with IMT • Request safety officer support • Request security officer support • Consider Continuity of Operations Plan (COOP) activation

4	**Regional emergency (low-level crisis)** Highly probable hazardous conditions and strong potential for property damage or loss of life The event may affect a limited portion of the region and require regional assistance.	• Single fatality • Significant long-term environmental effects • Severe long-term effects to critical infrastructure • Accidental spill or wastewater discharge violation with significant effect on facility, treatment plant, or river • \$100M–\$1B in damaged infrastructure • Serious public and media outcry • Significant fines and multiple litigations	• General manager or authorized representative declares an enterprisewide emergency • Affected utility facility(ies) are closed to all personnel except those designated as essential • Notify ECC—ECC will make incident level 4 notifications • Notify and identify the on-scene commander and/or incident commander • Full IMT activation by incident commander • Notify local HSEMA and make mandatory notifications to all local response agencies; notify all affected critical infrastructure personnel • Establish an incident command post and participate in the local/regional Joint Information Center/Joint Operations Center (JIC/JOC) if required • Conduct initial briefing with IMT and responders • Consider COOP activation

(continued)

TABLE 3.3 Response Matrix (*Continued*)

Level	Definition	Types of effects	Response actions
5	Regional emergency (high-level crisis) Extremely hazardous conditions that are imminent or occurring. The event affects the region and may require federal assistance.	• Multiple fatalities • Significant long-term environmental effects affecting regional ecosystems • Destruction of critical infrastructure • Accidental spill or wastewater discharge violation with severe effect on facility, treatment facility, or river • > $1 billion in damaged infrastructure • Prolonged international media condemnation • Potential jail terms for personnel and/or high fines with prolonged litigation	• General manager or authorized representative declares an enterprisewide emergency • Affected utility facility(ies) is closed to all personnel except those designated as essential • Notify ECC—ECC will make incident level 5 notifications • Notify and identify the on-scene commander and/or incident commander • Full IMT and COOP activation by incident commander • Notify local HSEMA and make mandatory notifications to all local/regional response agencies; notify all affected critical infrastructure personnel • Establish an incident command post and participate in the local/regional JIC/JOC if required • Conduct initial briefing with IMT and responders

TABLE 3.4 Typical Response Timeline

Activation and initial response procedure	
Activation	
• The general manager, on-scene commander, or previously designated facility incident commander may be the incident commander for purposes of activating the emergency management plan and establishing the incident management team. • The emergency management plan, with or without formation of an incident management team, may be activated only by the incident commander, who will be responsible for initiating the appropriate measures to inform and advise employees, contractors, and others of the appropriate safety measures to protect them during the emergency. • In general, the incident commander will base the activation decision on incident/threat level designations. • The incident commander should consult with the site safety and security officers, the general manager, the assistant general managers, and the emergency planning coordinator prior to making this decision, as time and circumstances permit. • Depending on the severity, effect, and, scope of the emergency, it is the role of the general manager to declare a "state of emergency." This declaration should be made in consultation with the incident commander and the director of safety and security and should be communicated to all employees and contractors.	
Within first 2 hours	
Task—First hour	**Responsible party**
• Determine need for 911 call • Call 911 if the incident threatens life, property, or the environment • Notify manager on site, supervisor, or general foreman • Become the on-scene commander until a senior employee, site manager, supervisor, or general foreman arrives	Employee who discovers the incident

(continued)

TABLE 3.4 Typical Response Timeline (*Continued*)

Within first 2 hours	
Task—First hour	**Responsible party**
Identify and designate the on-scene commander. Conduct initial incident assessment: • Determine need for 911 call • Assess whether incident is escalating or controlled • Determine resources needed • Identify incident potential to escalate	Site manager, supervisor, or general foreman
Conduct tactical operational response: • Conduct initial assessment • Identify incident escalation potential • Conduct onsite response—referring to incident-specific response plans and training • Conduct an initial situation briefing with staff	On-scene commander
If first responders (fire/police/other) have been notified and are on the site of the emergency, the on-scene commander will act as the incident liaison. Before arrival of 911 responders, the on-scene commander must be prepared to direct first responders to the exact location.	On-scene commander
Conduct appropriate notifications (including information contained in the Initial Incident Assessment Form) to the Emergency Command Center; executive management, including incident management team members; and key agency support.	On-scene commander or designee, or incident commander (if activated)

If a major injury or a death(s) has occurred to a utility employee or contract employee, immediate action must be taken to ensure the name(s) of the injured or deceased are not released to the press or outside the utility until the immediate family has been notified.	Incident commander Director of human resources Liaison officer General manager
• Confirm with the Emergency Command Center that the local homeland security and emergency management agency has been properly notified. There must be confirmation of any other mandatory notifications that need to be made.	On-scene commander, or incident commander (if activated)
• Conduct an emergency conference call with all utility personnel involved in the response. • Establish regular communications with the Emergency Command Center, assistant general manager(s) [(AGM(s)], general manager, liaison officer, and other departments involved in the response. The incident commander shall confirm the communications system has been established and is effectively providing regular communications with the emergency command center, liaison officer, and the incident management team (if activated).	On-scene commander, or incident commander (if activated), and coordinate with emergency planning coordinator
The on-scene commander designates a team to manage and abate the emergency.	On-scene commander
Deploy site safety and security officer(s) to the incident scene.	**Director of safety and security**
Task—Within 2 hours	Responsible party
Conduct an initial briefing as soon as time and circumstances permit, and hourly call-in briefings for all utility response personnel involved with the emergency.	Incident commander

(continued)

TABLE 3.4 Typical Response Timeline (*Continued*)

Task—First hour	**Responsible party**
Distribute the initial incident assessment form to the emergency management team (if activated), incident commander, general manager, all AGMs, director of safety and security, emergency planning coordinator, department directors, legal counsel, and other IMT members.	Emergency Command Center
Within first 24 hours	
Task—Preliminary summary report	
Within the first 24 hours, the on-scene commander will provide a written preliminary summary report to the general manager, liaison officer, legal officer, safety and security officer, emergency planning coordinator, and respective AGM. The report shall include the following: • A brief summary on what actually occurred; • Personnel involved (utility and/or contractor); • If injuries were involved, the status of the injured; • Effect(s) to utility operations (water, sewer system, wastewater treatment, or other services); • Timeline in which key actions occurred and/or corrective action used; and • Summary of the preliminary observations and findings.	
Task—Preliminary investigation	
The general manager is responsible for appointing a person responsible for conducting the preliminary investigation.	

Task—Final summary report
The emergency planning coordinator is responsible for the final summary report. The director of safety and security is responsible for providing information on any job-related accidents or injuries for this report. When the incident involves an operational emergency (probable legal or regulatory violations, equipment malfunctioning, etc.), the respective department head with oversight of the location of the emergency is responsible for providing this information for the report.
Within 10 business days
Task—Preliminary final written report
Within 10 business days of the emergency, the emergency planning coordinator is responsible for completing a preliminary final written report summarizing what occurred, why it occurred, actions taken to abate the situation, and recommendations for future corrective measures. The report shall be distributed to the members of the incident management team and others as determined by the general manager.
Within 30 days
Task—After-action report
Within 30 days of the emergency, the final written after-action report is due to the general manager. The emergency response manager will provide a written after-action report to be presented to senior management, other officials, the on-scene commander, and incident commander. The report will include the following: • Information taken from the initial incident assessment, • Incident briefing documents, and • Responder interviews. The report will describe the following: • Initial response actions; • Responsible parties; • Lessons learned; • Suggested after action-items; and • Updates to the current emergency management plan and when they will be implemented, and when they will be included in training.

TABLE 3.5 Planning During an Incident

1. Initial response period

The leg of the "P" describes the initial response period. Following are the steps taken once the incident/event begins:

- Notifications,
- Initial response and assessment,
- Incident briefing using ICS Form 201, and
- Initial incident command (IC)/unified command (UC) meeting.

2. First operational planning period

At the top of the leg of the "P" is the beginning of the first operational planning period cycle. Following are the steps in this circular sequence:

- IC/UC develop/update objectives meeting,
- Command and general staff meeting,
- Preparing for the tactics meeting,
- Tactics meeting,
- Preparing for the planning meeting,
- Planning meeting,
- Incident action plan prep and approval, and
- Operations briefing.

3. New operational planning period

At this point a new operational period begins. Following are the steps in this sequence:

- Execute plan,
- Assess progress, and
- Begin cycle again.

The ICS Form 201 provides command staff with information about the incident situation and the resources allocated to the incident. This form serves as a permanent record of the initial response to the incident and can be used for transfer of command.

3.3.1 *The Start of Each Planning Cycle*

An incident command/unified command objectives meeting establishes incident objectives that cover the entire course of the incident. For complex

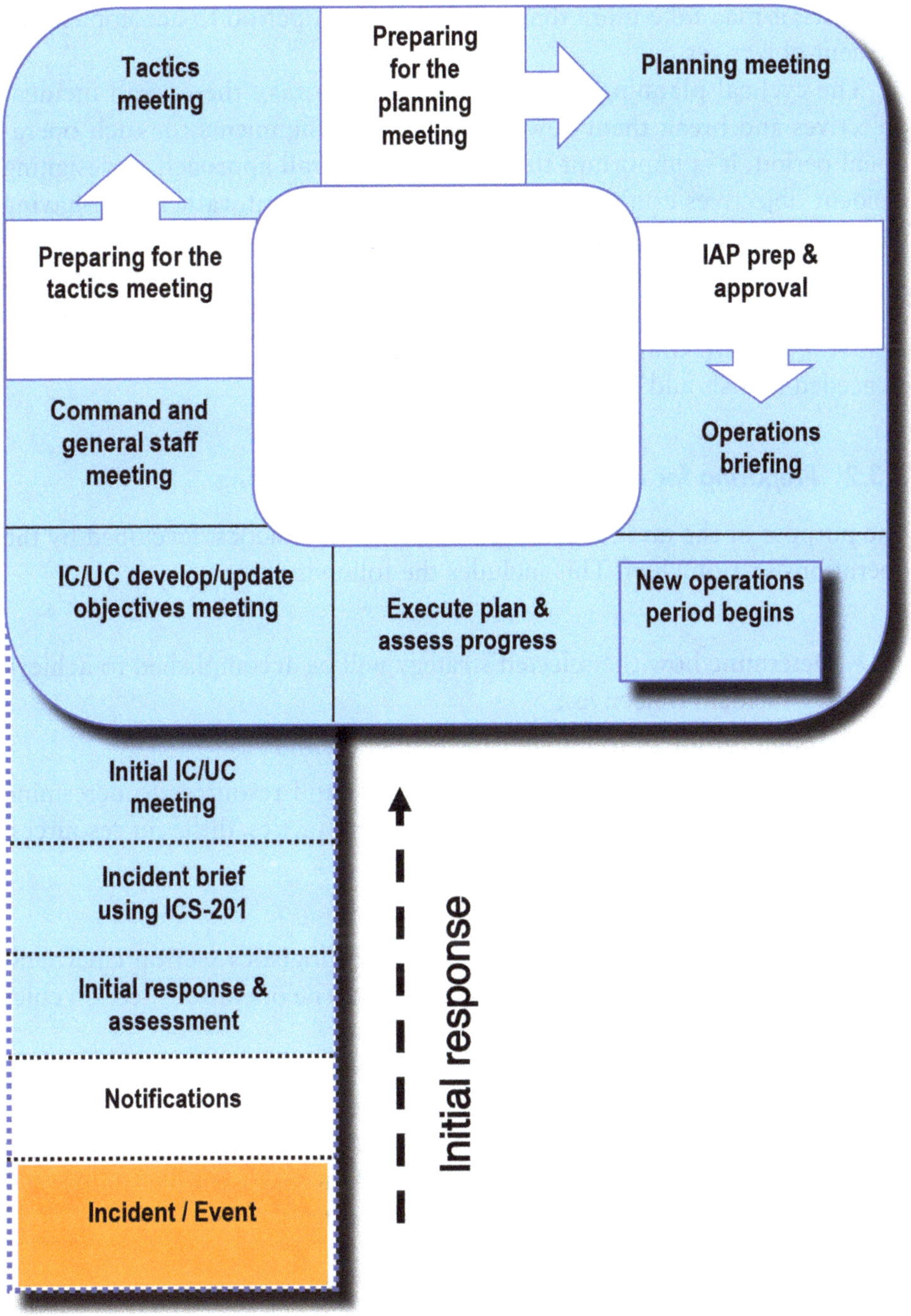

FIGURE 3.4 Planning Cycle

incidents, it may take more than one operational period to accomplish the incident objectives.

The cyclical planning process is designed to take the overall incident objectives and break them down into tactical assignments for each operational period. It is important that this initial overall approach to designing incident objectives establish the course of the incident, rather than having incident objectives that only address a single operational period.

The incident command/unified command may meet with the command and general staff to gather input or to provide immediate direction that cannot wait until the planning process is completed. This meeting occurs as needed and should be as brief as possible.

3.3.2 Preparing for and Conducting the Tactics Meeting

The purpose of the tactics meeting is to review the tactics developed by the operations section chief. This includes the following:

- Determine how the selected strategy will be accomplished to achieve the incident objectives;
- Assign resources to implement the tactics; and
- Identify methods for monitoring tactics and resources to determine if adjustments are required (e.g., different tactics, different resources, or new strategy).

The operations section chief, safety officer, logistics section chief, and resources unit leader attend the tactics meeting. The operations section chief leads the tactics meeting.

ICS Form 215, operational planning worksheet, and ICS Form 215A, incident safety analysis, are used to document the tactics meeting.

Resource assignments will be made for each of the specific tasks. Resource assignments will consist of the kind, type, and numbers of resources available and needed to achieve the tactical operations desired for the operational period.

If the required tactical resources will not be available, then an adjustment should be made to the tactical assignments being planned for the operational period. It is important that tactical resource availability and other needed support be determined before spending a great deal of time working on strategies and tactical operations that realistically cannot be achieved.

3.3.3 Preparing for the Planning Meeting

Following the tactics meeting, preparations are made for the planning meeting, to include the following actions coordinated by the planning section chief:

- Review ICS Form 215 developed in the tactics meeting;
- Review ICS Form 215A, incident safety analysis (prepared by the safety officer), based on the information in ICS Form 215;
- Assess current operations effectiveness and resource efficiency; and
- Gather information to support incident management decisions.

3.3.4 Conducting the Planning Meeting

The planning meeting provides the opportunity for the command and staff to review and validate the operational plan as proposed by the operations section chief. Attendance is required for all command and staff. Additional incident personnel may attend at the request of the planning section chief or the incident commander. The planning section chief conducts the planning meeting following a fixed agenda.

The operations section chief delineates the amount and type of resources required to accomplish the plan. The planning section's resources unit will work with the logistics section to accommodate its needs.

At the conclusion of the meeting, the planning section staff will indicate when all elements of the plan and support documents are required to be submitted so the plan can be collated, duplicated, and made ready for the operational period briefing.

3.3.5 Incident Action Plan Preparation and Approval

The next step in the incident action planning process is plan preparation and approval. The written plan is composed of a series of standard forms and supporting documents that convey the incident commander's intent and the operations section direction for accomplishing the plan for that operational period.

For simple incidents of short duration, the IAP will be developed by the incident commander and communicated to subordinates in a verbal briefing. The planning associated with this level of complexity does not demand the formal planning meeting process as highlighted in this guide.

Certain conditions result in the need for the incident commander to engage a more formal process. A written IAP should be considered in the following circumstance:

- Two or more jurisdictions are involved in the response,

- The incident continues into the next operational period,
- A number of ICS organizational elements are activated (typically when general staff sections are staffed),
- It is required by the affected utility's policy, and
- A hazardous materials incident is involved (required).

3.3.6 Operations Period Briefing

- The operations period briefing may be referred to as the operational briefing or the shift briefing. This briefing is conducted at the beginning of each operational period and presents the incident action plan to supervisors of tactical resources.
- After the operations period briefing, supervisors will meet with their assigned resources for a detailed briefing on their respective assignments.

3.3.7 Execute Plan and Assess Progress

- The operations section directs implementation of the plan. The supervisory personnel within the operations section are responsible for implementing the plan for the specific operational period.
- The plan is evaluated at various stages in its development and implementation. The operations section chief may make the appropriate adjustments during the operational period to ensure that the objectives are met and effectiveness is ensured.

3.4 Incident-Specific Procedures

The scope of this chapter does not cover details about developing and using incident-specific procedures. But it is important to identify the components of these procedures.

Most of such procedures should address key types of emergencies that may occur at any utility facility. Procedures can be generalized for potential hazards and detailed for specific incident response procedures when these are insufficient.

Hazards typically can be grouped under natural or technical/human-made categories. This list should be derived through a risk assessment process that reviews an extensive list of all risks and vulnerabilities. Table 3.6 lists examples of possible hazards and associated incidents/events. Table 3.7 outlines incident specific response procedures.

TABLE 3.6 Examples of Possible Hazards and Associated Incidents/Events

Natural (weather-related)	Human/technological hazards	
Atmospheric	**Biological**	**Disruption—civic**
Avalanches	Biological weapons of mass destruction (WMD)	Bomb/explosion
Cold/exposure	Critters	Civil unrest
Dust and sandstorms	Flu pandemic	Computer viruses
Extreme summer heat	Infectious disease	Cyberattacks
Global weather		Denial of services
Hailstorms		Email hoaxes
Hurricanes, cyclones		Hostage
Lightning		Mass hysteria/riot
Thunderstorms		Public violence
Tornadoes		Sabotage
Windstorms		Stand-off
Winter storms		Disruption—lone or few persons
		Lone gunman
		Unauthorized entry
		Vandalism
		Workplace violence/ disgruntled employee
Geologic	**Energy/utility incidents**	**Fires/explosions**
Coastal/rapid erosion	Cold/exposure	Arson
Droughts	Communication problems	Bomb/explosion
Expansive soils hydrologic	Extreme summer weather	Building
Floods and flash floods	Fuel shortage	Health care
Land subsidence/sinkholes	Power outage	Industrial
Landslides	Water shortage	Residential
Mudslides	Winter storms	
Storm surges		
Seismic and volcanic	**Hazardous materials**	**Medical**
Earthquakes	Asbestos	Fatality
Tsunamis	Chemical release or spills	Injury
Volcanoes	Hazardous materials/oil spills	Mass fatality
		Medical (such as heart attack)

(continued)

TABLE 3.6 (*Continued*)

Natural (weather-related)	Human/technological hazards	
Other natural disasters	**Terrorism**	**Transportation incidents**
Animal/insect infestation	Agro-terrorism	Accidents
Disease/biological (natural)	Cyberterrorism violence	Aviation related
Material from space	At home	Bridge collapse
Natural hazardous materials	At hospital	Incidents
Wildfire	At school	Ships, boats, barges & ferries
	At work	Surface vehicle
	Ecoterrorism	Train/subway crash
		Tunnel accident/fire
	Structural failure/accident	**Weapons of mass destruction**
	Bridge collapse	Biological WMD
	Building collapse	Chemical WMD
	Dam/levee failure	Radiological WMD
	Tunnel accident/fire	

TABLE 3.7 Incident-Specific Response Procedures

1. General

Define nature of threat and when it is likely to occur. Identify specific effects hazard may have. Identify likely source and timeframe of notification that hazard/threat will or could happen. Identify likely timing available for preparation and response.

Example: Flooding typically occurs during a heavy rainfall or snowmelt. Severe weather may include severe thunderstorms, high wind, lighting, heavy snowfall, and can result in inundation, soaking, high water forces that dislodge heavy objects, and landslides. There is typically sufficient time to take preventive measures if flooding or severe weather is imminent and the National Weather Service predictions are heeded.

2. Preparation

☐ Define who is mainly responsible for knowing what in preparing for possible threat. Define who monitors what, and what agency/organization is responsible for data/research/warnings.

Example: The emergency planning coordinator should be familiar with area flood and storm surge maps and other flood data. He or she monitors the National Oceanic and Atmospheric Administration's (NOAA's) National Weather Service (NWS) River Forecast Center (RFC). The RFC is responsible for issuing flood forecasts synthesized from hydrometerological data. Flash flood guidance is also issued. http://water.weather.gov/ahps/

(*continued*)

TABLE 3.7 (*Continued*)

☐ Define specific preparation responsibilities, such as what staff should know, have on hand, prepare for, and be trained for.

Example: Personnel that remain on site to operate the facility during a severe weather event should be prepared with proper weather gear for moving safely through the facility to secure and check on equipment. This gear includes winter wear and boots. Although the status of severe weather may eventually drop to steady rain or snow, for example, the initial winds marking the arrival of a storm can be extremely dangerous.

3. **Activation**

☐ Define when this response should be activated or be "on deck" for activation, and who is responsible for it.

Example: When hazardous weather conditions, many with potential flood hazards, are expected to affect the region, NWS issues public advisories. The emergency planning coordinator determines when and at what level response procedures should be initiated. He or she initiates response upon official notification by NWS of a "flood watch," "flash flood watch," or more elevated status, or if actual flooding is discovered.

4. **Notification/alarms**

☐ Define who notifies whom in the event of response activation, and what alarms are to be activated. This is the first step in any procedure. This section may reference the notification procedures elsewhere without duplicating them here. This may have multiple steps (such as notifying the emergency operations center, activating a localized alarm, initiating a facilitywide alarm, initiating evacuation).

5. **Response—incident command**

☐ Describe the exact response steps of leadership—incident commander—in this procedure. Reference any other procedures needed without duplicating them.

6. **Response—on-scene command—department or site-specific**

☐ Describe the response steps of on-scene commanders and other personnel by department, functional area, facility, or site. Area supervisors can become on-scene commanders. Reference any other procedure needed without duplicating them.

Describe process areas or system components that might be affected and how they are to be treated, including type and extent of effects of the emergency and response procedure.

7. **Response—other personnel/processes**

☐ Describe the response steps of any other personnel or processes as needed.

(*continued*)

TABLE 3.7 *(Continued)*

8. Return to service

- ☐ Describe all steps for return to service. A simple example follows.
  Example: The emergency planning coordinator is responsible for overseeing return to service. See Section X on Continuity of Operations.
- ☐ Contact state and local authorities to determine if there are any restrictions on disposal of materials and debris removed from the site or if a temporary discharge permit is needed (National Pollutant Discharge Elimination System or other).

9. Incident investigation and corrective action

- ☐ Describe postincident investigation steps.
  Example: The emergency planning coordinator conducts an incident investigation following the procedure in emergency management plan Section X. Appendix X contains the incident investigation form.
- ☐ The emergency planning coordinator, with the support of department and facility directors, files a corrective and preventive action (CAPA) plan as needed. CAPA includes a root-cause analysis. Appendix X contains the corrective and preventive action plan template.
- ☐ Assistant general managers, office chiefs, department directors, and facility directors are responsible for ensuring the CAPA plan is undertaken and completed for their respective jurisdictions. Section X addresses continuous improvement.
- ☐ The emergency planning coordinator ensures a review and update of the emergency management plan and related procedures, training, and documents are conducted based on lessons learned from the incident. Section X addresses continuous improvement.

4.0 INCORPORATING SAFETY TO DISASTER RESPONSE

Best practices ensure that all incidents in which an incident management team is activated must have a safety officer involved in the response. The site safety officer's role is to develop and recommend measures for ensuring personnel safety and to assess and/or anticipate hazardous and unsafe situations. Only one safety officer should be assigned for each incident.

Following are the typical responsibilities of the safety officer. Note that it is the responsibility of the incident commander to define the role of the safety officer for a particular incident, but the ICS has defined roles and responsibilities for all incident management team members, and those reporting to them.

- Report to incident scene when requested by the on-scene commander, incident commander.

- Ensure that all responders involved in the response have the appropriate personal protective equipment and that the equipment is used correctly.
- Identify hazardous or potentially unsafe situations associated with the incident and ensure that solutions are developed to correct the identified hazardous or potentially unsafe situations.
- Prepare the incident safety plan (ICS Form 208, see below). Distribute the plan to all personnel involved in the response.
- Maintain a log recording all safety-related events and decisions.
- Obtain a situation briefing from the on-scene commander or incident commander (if activated).
- Review any response actions for potential safety implications.
- Exercise emergency authority to stop and prevent unsafe acts.
- Keep the on-scene commander or incident commander (if activated) informed of operational safety problems and potential hazards.
- Provide a safety report and incident critique to emergency planning coordinator.

Figure 3.5 is ICS 208, a procedure for creating and disseminating a safety message. This is one of a set of ICS forms and plans available for customizing by any organization.

5.0 COMMUNICATION: PLANNING BECOMES DOING

As part of preemergency planning, every agency should have developed an emergency response notification and communication flowchart with up-to-date contact information for critical personnel. Roles and responsibilities of the public information officer and communication team should be clear as well.

Information for a response procedure can be adapted from the CDC *Crisis Emergency Risk Communication Manual* (http://emergency.cdc.gov/cerc/) and the *Crisis & Emergency Risk Communication Tool Kit for Community Water Systems*. Once a crisis hits, everything changes from planning to doing. Be prepared to be peppered with questions when the media discover the crisis, and not much may have been determined than what the media already know.

Many crises begin with a telephone call from a reporter. Get all the information possible from the reporter, including how he or she found out

ICS 208
Safety Message/Plan

Purpose. The Safety Message/Plan (ICS 208) expands on the Safety Message and Site Safety Plan.

Preparation. The ICS 208 is an optional form that may be included and completed by the Safety Officer for the Incident Action Plan (IAP).

Distribution. The ICS 208, if developed, will be reproduced with the IAP and given to all recipients as part of the IAP. All completed original forms must be given to the Documentation Unit.

Notes:
- The ICS 208 may serve (optionally) as part of the IAP.
- Use additional copies for continuation sheets as needed, and indicate pagination as used.

Block Number	Block Title	Instructions
1	**Incident Name**	Enter the name assigned to the incident.
2	**Operational Period** • Date and Time From • Date and Time To	Enter the start date (month/day/year) and time (using the 24-hour clock) and end date and time for the operational period to which the form applies.
3	**Safety Message/Expanded Safety Message, Safety Plan, Site Safety Plan**	Enter clear, concise statements for safety message(s), priorities, and key command emphasis/decisions/directions. Enter information such as known safety hazards and specific precautions to be observed during this operational period. If needed, additional safety message(s) should be referenced and attached.
4	**Site Safety Plan Required?** Yes ☐ No ☐	Check whether or not a site safety plan is required for this incident.
	Approved Site Safety Plan(s) Located At	Enter where the approved Site Safety Plan(s) is located.
5	**Prepared by** • Name • Position/Title • Signature • Date/Time	Enter the name, ICS position, and signature of the person preparing the form. Enter date (month/day/year) and time prepared (24-hour clock).

FIGURE 3.5 Safety Message/Plan

about the situation. Was it a call from a concerned citizen? From police? From another utility? All information helps at this stage.

5.1 Say What Can Be Said

Tell the reporter the matter will be investigated and that he or she will be called back as soon as the facts are determined. Afterward, call back with what is called a "holding statement."

5.2 Create a Holding Statement

A holding statement refers back to the precipitating event, often confirming that something has happened and you know about it. Its purpose is to buy the organization time during which the person who answered the telephone inquiry and the crisis management team begin to dissect the problem and get solutions started. If the event is a significant one that could endanger lives or cause societal disruptions, also tell the reporter what people need

SAFETY MESSAGE/PLAN (ICS 208)

1. Incident Name:	**2. Operational Period:** Date From: Date To: Time From: Time To:
3. Safety Message/Expanded Safety Message, Safety Plan, Site Safety Plan:	
4. Site Safety Plan Required? Yes ☐ No ☐ **Approved Site Safety Plan(s) Located At:**	
5. Prepared by: Name: ______________	Position/Title: ______________ Signature: ______________
ICS 208 | **IAP Page** _____	Date/Time: ______________

FIGURE 3.5 (*Continued*)

to do to be safe and that the information will be released to other media outlets as well because the public needs to know.

5.3 Tell More of What Can Be Said

It is important to communicate after the problem is determined. The danger from not communicating is that someone else will fill the void if the spokesperson does not. In today's world of cell phones, cameras, YouTube, Facebook, and Twitter, the organization's event can be all over the news minutes after something happens.

When U.S. Airways Airbus A320 jet crashed in the Hudson River in New York in January 2009, the first news of the event was transmitted using Twitter. A London (U.K.) newspaper, the *Telegraph*, said: "Twitter users broke the news of the incident around 15 minutes before the mainstream media alerted viewers and readers to the crash. The first recorded tweet about the crash came from Jim Hanrahan, aka Manolantern, four minutes after the plane went down, who wrote: 'I just watched a plane crash into the hudson riv [sic] in manhattan'" (Beaumont, 2009).

The Public Relations Society of America's monthly newspaper, *Tactics,* said that not only did the first information come from Twitter, "the first photo the world saw of it came through Twitter. After a significant event, the first news will often come through social media" (Baron and Philbin, 2009).

5.4 Prepare for Two-Way Conversations; Social Media Can Be a Two-Edged Sword

Conversation, if used properly, can be a good way to supplement communication tools to communicate about a crisis and what the organization is doing to resolve it. Be prepared to have legitimate two-way conversations (not just push information out). The organization must trust the spokesperson to engage in these conversations. Before engaging, make sure the organization has developed a good social media policy and employees know that only the spokesperson speaks for the organization.

5.5 Express Concern for Victims

If loss of life, injury, or significant property damage to others' property or the environment has occurred, the holding statement should first express concern for victims. Facts are not enough. Also respond to people's emotional needs. Be compassionate, confident, and competent.

A brief who-what-when-where-how-and-why statement should follow the expression of concern. Get out all the information possible, including information that might feel embarrassing or painful. Often, in a crisis, not much is known. The first holding statement can simply indicate concern about the victims, acknowledging something significant has happened, and that the organization is working on a solution. Subsequent statements can then be updates. If possible, conclude with information about when the next update can be expected. If a "next update" time is published, expect the media to follow up to make sure they do not miss something important.

5.6 Respond to Questions the Media Will Ask

Many of the questions the media will ask are predictable. U.S. EPA document titled *Effective Risk and Crisis Communication during Water Security Emergencies* (EPA/600/R-07/027) contains a list of "The 77 Most Frequently Asked Questions by Media Following Crisis Incidents." A few sample questions are listed below:

1. What is your name and title?
2. What are your job responsibilities?
3. What are your job qualifications?
4. Can you tell us what happened?
5. When did it happen?
6. Where did it happen?
7. Who was harmed?
8. Are those that were harmed getting help?
9. How certain are you about this information?
10. How are those who were harmed getting help? (Covello et al., 2007)

Use the list to help the organization prepare in developing answers.

5.7 Avoid Saying "No Comment"

Do not say, "No comment." There are hundreds of ways not to comment on an event without saying the words "No comment." Media training can help the spokesperson learn some of these ahead of time. Early in an unfolding situation, for example, a simple "I do not know but I will check and get back to you" response can allow time to get back in control of the situation. If a return call is promised, however, it must be done. Otherwise, the organization's reputation may be affected.

5.8 Identify Affected Parties

With omnipresent camera-equipped cell phones and smart phones, nothing significant will remain unpublicized for long. The first priority is to limit or end the problem (if possible) as quickly as possible; then the information must be communicated. Think in terms of stakeholders:

- Those affected (victims, their families, other stakeholders);
- Employees;
- Government, regulators, neighbors, community leaders, stakeholders; and
- News media and other external channels of communication.

Victims, their families, and other stakeholders may not be known ahead of time, but many of those noted above are known to the organization. Names, titles, telephone numbers, and email addresses; regulators; community leaders; and other stakeholders can be listed in the communication plan ahead of time so they are easy to find when a crisis occurs. U.S. EPA's document titled *Effective Risk and Crisis Communication during Water Security Emergencies* (EPA/600/R-07/027) contains a long list of potential stakeholders (Covello et al., 2007).

Employees are an especially important stakeholder group. They are not only involved and part of the team, but when they go home they will be quizzed by their families and friends. Nobody likes to feel ignorant when being questioned by people close to them, so be sure to make employees feel like trusted insiders, even if it is only: "We do not know yet what caused this, but here is what I can tell you."

5.9 Prepare Written Statements

As the event unfolds, it may be advisable to communicate with the media via written statements. These bare-bones, single-issue communications can be issued via email to the organization's media list and posted on the organization's website. They provide a written record and to keep all the information sequential, clear, and accurate. Statements can contain quotes from senior officials, which add perspective. A website or a blog site can be set up specifically for the crisis. On it statements, background information, photos, diagrams, graphics, and even videos can be posted. Sequential statements also help as the event grows and more media representatives become involved because they can be directed to the website or blog for background.

Eventually, the media (especially television) will begin to clamor for a live media briefing. At some point, it may be advisable to have the spokesperson hold a media briefing or even a press conference (if the event warrants it). A briefing is typically easier to conduct. It does not carry with it the media's expectations for a comprehensive question-and-answer period that a press conference does. As noted above in Section 3.4, many of the questions can be anticipated.

5.10 Conduct a Media Briefing

A media briefing without an opportunity for media members to ask questions is basically a statement offered to the cameras by a live spokesperson. The conditions for such a briefing (time, location, rules of engagement) must be spelled out clearly in the advance notification of the briefing that is sent to the media the media's expectations can be managed. If an answer to a question is not expected, say so. A full-blown press conference is more difficult to plan and manage properly, carries higher expectations (including an extensive question-and-answer session with a senior official), and should be done only with the help of an experienced media relations team and trained senior officials.

If the organization chooses to hold a media briefing, begin it with an expression of concern for victims (if there are victims), even that concern was already expressed in the written statements. Begin the statement with a short, factual who-what-when-where-why-and-how statement that updates the situation and recaps what is known. Use short, declarative sentences. Try to keep the whole statement to 20 seconds. A complex situation could run 2 or 3 minutes, but keep it short. The media will extract the most dramatic 10 seconds.

Outline what steps will be and have been taken to resolve the emergency. Have someone videotape the statement. Remember: the statement should talk directly to the stakeholders, not the media. If the intent is to take questions, do so now. Otherwise, sit down and be quiet. Back up the live media briefing statement with handouts and a written copy posted on the organization's website. Visuals (such as a time line) at the briefing site and on the website can help.

5.11 Take Photographs

Photographs can help document what things looked like during the beginning, middle, and end of the event. Be sure to take (or have someone else take) photographs as the event unfolds. After determining that it is safe enough to allow media representatives to a site to shoot photographs or video footage, make sure each cameraperson and photographer has an individual escort. Ensure that the escorts know where the cameras can and cannot go. Safety is the primary concern. The organization is responsible because media people do not know the facility or working conditions. Media members can easily get into dangerous situations or get hurt when they are focused on getting the defining photograph or video clip of the event. Remember also that media can hire helicopters as photograph platforms, so do not believe that an embarrassing situation can be kept hidden from view.

An alternative to letting a large group of media onto the site for pictures is to allow a pool still photographer (representing the print media) and a pool videographer (representing TELEVISION) onto the site under escort. The rules for a pool photographer and videographer are that they share what they shoot equally with all of the other media. If a pool photographer and videographer are allowed, let the media choose from among themselves who these media representatives should be. That way, the organization of spokesperson cannot be accused of playing favorites.

5.12 Additional Tips

Some additional useful tips for managing the communication portion of a crisis include

- Treat all media equally (do not give information to one and not the others);
- Control the scene (to ensure their safety);
- Be consistent in the story (this is why it is important to tell the truth);
- Tell the truth (see "be consistent" above);
- Do not speculate (deal only with what has happened/is happening, not what might have happened or might yet happen, and speak only about what is known);
- Nothing is off the record (always consider the microphone to be live and the camera rolling);
- Cooperate with law enforcement and regulators (they may be questioned by the media);
- Remind employees not to talk to media (there must be only one spokesperson);
- Monitor the media coverage so inaccuracies can be corrected; and
- Monitor social media and the Internet to hear what people are saying and what concerns them.

Not all emergencies or crises are of long duration. Some are over in a day, but others can extend over several days or longer. Eventually, however, the period of acute crisis or emergency will end. Things will appear to return to normal. That is when the organization must do its postevent analysis to make sure it understands what happened, make sure the responses were appropriate, and make sure the crisis does not recur. This is when the stakeholders can be shown that the crisis was an anomaly.

For some time, the media and others may want to revisit the crisis every time they talk about the organization. This is natural. Eventually, however, it will become old news and they will tire of it. Meanwhile, review what happened, revise plans accordingly, and practice with the internal team for the next event. That will enable the organization to be ready to show the media and stakeholders what the organization does and the service it renders to the community. Many people, especially visitors, will find what the clean water community does is amazing and will go home extolling its virtues. This is the time to put goodwill credits in the public opinion bank for the next time they are needed.

6.0 RESOURCES

Table 3.8 lists resources and best practices. Table 3.9 includes industry standards, tools, and best practices.

TABLE 3.8 Resource List and Best Practices

Government standards/best practices		
Organization	**Website**	**Standard/Practice description**
Centers for Disease Control and Prevention (CDC)		
CDC Drinking Water Advisory Communication Toolbox	http://www.cdc.gov/healthywater/emergency/drinking_water_advisory/index.html	This toolbox provides a protocol and practical guide for communicating with stakeholders and the public about water advisories that is based on research and identified practices. The toolbox focuses on water systems and addresses the spectrum of situations that generate drinking water advisories. This project was a collaborative effort among the CDC, U. S. Environmental Protection Agency (U. S. EPA), American Water Works Association (AWWA), Association of State and Territorial Health Officials (ASTHO), Association of State Drinking Water Administrators (ASDWA), and National Environmental Health Association (NEHA).
Federal Bureau of Investigation		
Weapons of Mass Destruction Directorate	http://www.fbi.gov/about-us/investigate/terrorism/wmd	Part of a robust program to address terrorism, including weapons of mass destruction (WMD) and criminal use of chemicals. This website and related documents and programs include guidance on preparing for and responding to WMD and chemical attacks. There is a local WMD coordinator for each of its 56 field offices who will work with local authorities on reporting and responding to threats. The FBI also conducts workshops on water terrorism.

National Environmental Services Center (NESC), West Virginia University		
Clearinghouse; NESC services; funding from U.S. EPA	http://www.nesc.wvu.edu/	Information about protecting small drinking water and wastewater systems against potential threats. Includes a collection of articles, checklists, telephone numbers, and Web sites from NESC and other organizations across the country.
National Integrated Drought Information System (NIDIS)		
National drought warning and information system	http://www.drought.gov/drought/	NIDIS provides users with the ability to determine the potential effects of drought and the associated risks they bring, and the decision support tools needed to better prepare for and mitigate the effects of drought. In this, NIDIS forms the backbone of a national Drought Early Warning System and the U.S. Drought Portal is the public face of NIDIS on the Web.
U.S. Coast Guard (USCG)		
National Search and Rescue Committee, USCG	http://www.uscg.mil/hq/cg5/cg534/NSARC.asp	The National Search and Rescue Committee is a federal-level committee formed to coordinate civil search and rescue (SAR) matters of interagency interest within the United States.

(*continued*)

TABLE 3.8 Resource List and Best Practices (*Continued*)

Government standards/best practices		
Organization	**Website**	**Standard/Practice description**
U.S. Department of Homeland Security (DHS)		
Federal Emergency Management Agency (FEMA); National Incident Management System (NIMS)	http://www.fema.gov/emergency/nims/	NIMS provides a systematic, proactive approach to guide departments and agencies at all levels of government, nongovernmental organizations, and the private sector to work seamlessly to prevent, protect against, respond to, recover from, and mitigate the effects of incidents, regardless of cause, size, location, or complexity, to reduce the loss of life and property and harm to the environment. NIMS works hand in hand with the National Response Framework (NRF). NIMS provides the template for management of incidents, while the NRF provides the structure and mechanisms for national-level policy for incident management.
FEMA Emergency Management Guide for Business and Industry	http://www.fema.gov/business/guide/toc.shtm	An emergency management guide for business and industry that can be used by manufacturers, corporate offices, retailers, utilities, or any organization where a sizable number of people work or gather. Includes planning, emergency considerations, hazard-specific information, and information sources.

FBI *Cyber crime*	https://www.fbi.gov/investigate/cyber	Established by the FBI's Cyber Division in 2006 to provide rapid incident response on major computer intrusions and cyber-related emergencies, the team has approximately 50 members located in field offices around the country. They are either special agents or computer scientists, and all possess advanced training in computer languages, forensic investigations, and malware analysis. And since the team's inception, the Bureau has investigated hundreds of cyber crimes, and a number of those cases were deemed of such significance that the rapid response and specialized skills of the Cyber Action Team were required. Some of those cases affected U.S. interests abroad, and the team deployed overseas, working through our legal attaché offices and with our international partners.
National Preparedness Guidelines	Guidelines: http://www.dhs.gov/files/publications/gc_1189788256647.shtm Universal Task List: www.fema.gov/pdf/government/training/tcl.pdf.	The purposes of the guidelines are to • Organize and synchronize national (including federal, state, local, tribal, and territorial) efforts to strengthen national preparedness; • Guide national investments in national preparedness; • Incorporate lessons learned from past disasters into national preparedness priorities; • Facilitate a capability-based and risk-based investment planning process; and • Establish readiness metrics to measure progress and a system for assessing the nation's overall preparedness capability to respond to significant events, especially those involving acts of terrorism.

(*continued*)

TABLE 3.8 Resource List and Best Practices (*Continued*)

Government standards/best practices		
Organization	**Website**	**Standard/Practice description**
DHS Infrastructure Protection Plan (NIPP)	http://www.dhs.gov/files/programs/editorial_0827.shtm	This integrates the nation's critical infrastructure and key resources (CIKR) into a single national program. First released in 2006, the revised NIPP integrates the concepts of resilience and protection, and broadens the focus of NIPP-related programs and activities to an all-hazards environment. The U.S. Department of Homeland Security oversees NIPP management and implementation.
Homeland Security Exercise and Evaluation Program (HSEEP)	https://hseep.dhs.gov/pages/1001_HSEEP7.aspx	The HSEEP is a capabilities and performance-based exercise program that provides a standardized methodology and terminology for exercise design, development, conduct, evaluation, and improvement planning. The HSEEP constitutes a national standard for all exercises. Through exercises, the National Exercise Program supports organizations to achieve objective assessments of their capabilities so that strengths and areas for improvement are identified, corrected, and shared as appropriate before a real incident The HSEEP is maintained by FEMA's National Preparedness Directorate, Department of Homeland Security.

NIMS Standards Development (DHS-maintained national standard list)	http://www.fema.gov/national-incident-management-system/ongoing-management-and-maintenance	The National Integration Center works with appropriate standards development organizations to ensure the adoption of common national standards and credentialing systems that are compatible and aligned with the implementation of NIMS.
Fusion centers	http://www.dhs.gov/state-and-major-urban-area-fusion-centers	These are regional focal points to centralize threat-related information for planning and in real-time during events. They receive, analyze, gather, and share information among all levels of government and private-sector partners and stakeholders.
Protective security advisors (PSAs)	http://www.dhs.gov/protective-security-advisors	PSAs are trained experts in critical infrastructure protection and vulnerability mitigation. They conduct specialized site visits and can work with utilities to provide guidance on protecting critical infrastructures.

(continued)

TABLE 3.8 Resource List and Best Practices (*Continued*)

Government standards/best practices		
Organization	**Website**	**Standard/Practice description**
Hazardous Response Program	http://www.dhs.gov/hazardous-response-program	This Federal Protective Service (FRP) program provides response to credible chemical, biological, radiological, nuclear, and high-yield explosive (CBRNE) threats or incidents. It conducts initial investigations of suspicious or threatening CBRNE incidents and complies with OSHA and NFPA guidance and regulations. Response teams are available to utilities.
FEMA flood maps	http://www.floodmaps.fema.gov/	Information and maps of area floodplains.

U.S. Department of Veterans Affairs, National Center for Post-Traumatic Stress Syndrome (PTSD)		
Psychological First Aid: Field Operations Guide, September, 2nd Edition, 2006	http://www.ptsd.va.gov/professional/manuals/psych-first-aid.asp	Developed jointly with the National Child Traumatic Stress Network, psychological first aid is an evidence-informed modular approach for assisting people in the immediate aftermath of disaster and terrorism: to reduce initial distress, and to foster short- and long-term adaptive functioning. It is for use by first responders, incident command systems, primary and emergency health care providers, school crisis response teams, faith-based organizations, disaster relief organizations, CERTs, Medical Reserve Corps, and the Citizens Corps in diverse settings. Includes handouts for survivors.
U.S. Environmental Protection Agency (U.S. EPA)		
Water Security Publications Database	http://cfpub.epa.gov/safewater/watersecurity/publications.cfm	Guidance, documents, reports and studies, training (including NIMS for water sector).
Vulnerability Self-Assessment Tool (VSAT)	http://water.epa.gov/infrastructure/watersecurity/techtools/	VSAT is a risk-assessment software tool for water, wastewater, and combined utilities of all sizes. The tool assists drinking water and wastewater facility owners and operators in performing security threats and natural hazards risk assessments, as well as updating utility emergency response plans. U.S. EPA, in collaboration with water sector partners and stakeholders, updated VSAT to be consistent with the risk analysis and management framework for critical asset protection. The tool includes a user interface, an enhanced natural disaster threat assessment process, and a revised risk assessment approach.

(*continued*)

TABLE 3.8 Resource List and Best Practices (*Continued*)

Government standards/best practices		
Organization	**Website**	**Standard/Practice description**
Key Features of an Active and Effective Protective Program Tools and Resources To Assist Utilities	http://water.epa.gov/infrastructure/watersecurity/features/upload/epa817k12003.pdf	A list of tools and resources utilities can use to implement key feature—basic guidelines—for developing water utility protective programs that are flexible enough for utilities of all sizes and locations, and that are consistent with the management philosophy of continuous improvement.
Climate Resilience Evaluation and Awareness Tool (CREAT)	http://water.epa.gov/infrastructure/watersecurity/climate/creat.cfm	This software tool helps water sector owners and operators understand potential climate change threats and assess related risks. Climate change has been linked to increasingly extreme, frequent, and unpredictable weather events. While this tool is more in line with long-term planning, putting it to use can mitigate emergencies and reduce the need for extreme response efforts.
Federal Funding for Utilities—Water/Wastewater—in National Disasters (Fed FUNDS)	http://water.epa.gov/infrastructure/watersecurity/funding/fedfunds/index.cfm	Fed FUNDS provides tailored information to water and wastewater utilities about applicable federal disaster funding programs. Much of this pertains to FEMA funding. The Fed FUNDS Web pages address national-level disasters, but could apply to large-scale and local disasters that result in service interruptions and significant damage to the critical water/wastewater infrastructure. Four sections relate to pre-disaster, during a disaster, postdisaster and utility examples, training and assistance support. The last contains examples of how others have obtained funding; learn from others how the funds can be used. A training Webinar was planned by U.S. EPA.

U.S. Occupational, Safety, and Health Administration (OSHA)		
Personal protective equipment (PPE) standards	http://www.osha.gov/SLTC/personalprotectiveequipment/	Highlights OSHA standards, preambles to final rules (background to final rules), Federal Registers (rules, proposed rules, and notices), directives (instructions for compliance officers), standard interpretations (official letters of interpretation of the standards), and national consensus standards related to PPE.
Sandia National Laboratory		
Sandia National Laboratories Security Risk Assessment Methodologies	www.sandia.gov/ram	Performance-based methods for designing and evaluating physical protection systems (PPS). Provides risk assessment methodology (RAM) based on evaluation and design of an integrated performance-based system. Many utilities follow the RAM refined by Sandia National Laboratory.

TABLE 3.9 Industry Standards, Tools, and Best Practices

Organization	Website	Standard/Practice description
American National Standards Institute (ANSI)		
INCITS 398-2008 Information Technology —Common Biometric Exchange Formats Framework (CBEFF)	http://webstore.ansi.org/RecordDetail.aspx?sku=ANSI+INCITS+398-2008#.UHyzAWl27ms	This standard (revision of ANSI INCITS 398-2005) specifies a common set of data elements necessary to support multiple biometric technologies and to promote interoperability of biometric-based application programs and systems by allowing for biometric data exchange. These common data elements can be placed in a single file, record, or data object to exchange biometric information between different system components and applications. This standard specifies the biometric data elements.
American Society of Civil Engineers		
Guidelines for the Physical Security of Wastewater/Stormwater Utilities	http://www.awwa.org/files/science/WISE/Guidelines_Physical_Security_Wastewater_Stormwater.pdf	This Draft American National Standard for Trial Use (DSTU) applies to physical security for wastewater collection and treatment systems (also called sanitary sewer collection and treatment systems) and stormwater systems. These wastewater/stormwater utilities guidelines recommend physical and electronic security measures for physical protection systems to protect against identified adversaries referred to as the design basis threats (DBTs), with specified motivation, tools, equipment, and weapons. These guidelines describe physical security approaches to delay or detect malevolent threats.

Other associations/organizations		
Water and Wastewater Agency Response Networks (WARN)	http://apps.awwa.org/ebusmain/warn.aspx	A network of utilities helping other utilities to respond to and recover from emergencies. The purpose of a WARN is to provide a method whereby water/wastewater utilities that have sustained or anticipate damages from natural or human-caused incidents can provide and receive emergency aid and assistance in the form of personnel, equipment, materials, and other associated services as necessary from other water/wastewater utilities.
Water Infrastructure Security Enhancements	http://www.awwa.org/Resources/Content.cfm?ItemNumber=29824&navItemNumber=29837	The American Society of Civil Engineers (ASCE), the American Water Works Association (AWWA), and the Water Environment Federation (WEF), with a grant from the U.S. Environmental Protection Agency (U.S. EPA), implemented a three-phased program to address physical infrastructure security needs for water supply, wastewater and stormwater, and online contaminant monitoring systems. Includes guidance documents and training materials.

(*continued*)

TABLE 3.9 Industry Standards, Tools, and Best Practices (*Continued*)

Organization	Website	Standard/Practice description
ANSI/ASME-ITI/AWWA J100 RAMCAP		
Standard for Risk and Resilience Management of Water and Wastewater Systems	http://www.awwa.org/Resources/standards.cfm?ItemNumber=54453&navItemNumber=55050	Joint standard: American National Standards Institute (www.ansi.org), American Water Works Association (www.awwa.org), and American Society of Mechanical Engineers (ASME) Innovative Technologies Institute (www.asme-iti.org). Voluntary consensus standard encompassing an all-hazards risk and resilience management process for use specifically by water and wastewater utilities. The standard received final clearance from the American National Standards Institute (ANSI) in May 2010 and was effective July 1, 2010.
Institute of Electrical and Electronics Engineers (IEEE)		
Emergency Management Assistance Compact (EMAC)	http://www.iteris.com/itsarch/html/standard/ieeeim.htm	The incident management family of standards, created primarily by the IEEE standards development organization, addresses the interfaces between an emergency management center and other centers. They provide the vocabulary (called data elements and messages) necessary to exchange information between intelligent transportation systems. Together, the ITS standards in this group apply to emergency management center interfaces (such as with a traffic management center, an emergency management center, or other centers). There is also a significant amount of reuse of data elements and messages across multiple interfaces; rather than repeat the entire list of standards for each architecture flow, we have created this summary entry—the Incident Management (IEEE IM) Group of standards.

National Environmental Services Center (NESC), West Virginia University		
Clearinghouse; NESC services; funding from U.S. EPA	http://www.nesc.wvu.edu/	Information about protecting small drinking water and wastewater systems against potential threats. Includes a collection of articles, checklists, telephone numbers, and websites from NESC and other organizations across the country.
National Emergency Management Association (NEMA)		
Emergency Management Assistance Compact (EMAC)	http://www.emacweb.org/	EMAC is the first national disaster–relief compact since the Civil Defense and Disaster Compact of 1950 to be ratified by Congress. Through EMAC, a disaster-affected state can request and receive assistance from other member states quickly and efficiently, resolving two key issues upfront: liability and reimbursement. The National Emergency Management Association (NEMA) is a nonpartisan, nonprofit 501(c)(3) association dedicated to enhancing public safety by improving the nation's ability to prepare for, respond to, and recover from all emergencies, disasters, and threats to our nation's security.

(continued)

TABLE 3.9 Industry Standards, Tools, and Best Practices (*Continued*)

Organization	Website	Standard/Practice description
National Fire Protection Association (NFPA)		
NFPA 1600 (Disaster Emergency Management and Business Continuity Programs) and NFPA 1561 (Emergency Services Incident Management Systems)	http://www.nfpa.org/assets/files/pdf/nfpa1600.pdf	This (free) standard establishes a common set of criteria for disaster/emergency management and business continuity programs. It provides disaster and emergency management and business continuity programs, the criteria to assess current programs or to develop, implement, and maintain aspects for prevention, mitigation, preparation, response, and recovery from emergencies. This standard includes valuable appendices with dozens of resources and organizations that support emergency management activities.
NFPA 1221: Standard for Installation, Maintenance, and Use of Emergency Services Communications Systems	http://www.nfpa.org/assets/files/PDF/CodesStandards/TIAErrataFI/TIA1221-02-1.pdf	Reliable emergency communications are an integral part of effective emergency responsiveness by the fire service, law enforcement, emergency medical services, and other responders. This standard provides procedural and physical criteria for the installation, performance, operation, and maintenance of public emergency service communications systems and facilities.

Organization for the Advancement of Structured Information Standards (OASIS)		
https://www.oasis-open.org/committees/download.php/6334/oasis-200402-cap-core-1.0.pdf	https://www.oasis-open.org/committees/download.php/6334/oasis-200402-cap-core-1.0.pdf	Adopted by NIMS, the Common Alerting Protocol (CAP) is a simple but general format for exchanging all-hazard emergency alerts and public warnings over all kinds of networks. CAP allows a consistent warning message to be disseminated simultaneously over many different warning systems, thus increasing warning effectiveness while simplifying the warning task. CAP also facilitates the detection of emerging patterns in local warnings of various kinds, such as might indicate an undetected hazard or hostile act. CAP provides a template for effective warning messages based on best practices identified in academic research and real-world experience.
Pan American Health Organization Regional Office of the World Health Organization		
Natural Disaster Mitigation in Drinking Water and Sewerage Systems Guidelines for Vulnerability Analysis	http://www.paho.org/english/Ped/nd-water_mit.pdf	Provides basic tools water services can use to evaluate the components of their systems that are vulnerable to major natural hazards (earthquakes, hurricanes, floods, landslides, volcanic eruptions, and drought).

(*continued*)

TABLE 3.9 Industry Standards, Tools, and Best Practices (*Continued*)

Organization	Website	Standard/Practice description
Water Environment Federation		
Water Security & Emergency Response	http://www.wef.org/AWK/pages_cs.aspx?id=640	Contains numerous updated documents and links to tools, resources, and technical papers related to water/wastewater industry and government practices, procedures, and technology.
Water Environment Research Foundation		
03-CTS-4S, Emergency Response Plan Guidance for Wastewater Systems	http://www.werf.org/AM/Template.cfm?Section=Research_Profile&Template=/CustomSource/Research/PublicationProfile.cfm&id=03-CTS-4S	Guides wastewater facilities in the development of emergency response plans for natural or humanmade (i.e., terrorist-related) events, and presents a reference for the information and data that should be included in an emergency response plan. Additionally, this report illustrates important emergency planning concepts in a wastewater system context, such as chain-of-command, communication and notification plans, personnel safety provisions, plan initiation and decision-making process, emergency operations centers, and emergency response training.

Case studies and other examples and resources		
The Water Information Sharing and Analysis Center (WaterISAC)	https://portal.waterisac.org/web/	WaterISAC was authorized by Congress in 2002 to keep drinking water and wastewater utility managers informed of potential risks to the nation's water infrastructure from contamination, terrorism, and cyber threats. It is created and managed by the water sector and links members through a secure online portal.
Homeland Security Information Network (HSIN)	http://www.dhs.gov/homeland-security-information-network	This brings together emergency services professionals in "communities of interest" (COIs), and provides situational awareness tools.
Community Emergency Response Teams (CERT) Program	http://www.awwa.org/resources-tools/water-knowledge/emergency-preparedness/water-wastewater-agency-response-network.aspx	This educates people about disaster preparedness for hazards that may affect their area and helps neighborhoods and localities work and respond together.

(*continued*)

TABLE 3.9 Industry Standards, Tools, and Best Practices (*Continued*)

Organization	Website	Standard/Practice description
Multiagency Response to a Major Water Pipe Break: A Massachusetts Case Study and Evaluation	https://portal.waterisac.org/web/MWRA_web.pdf	Water utilities are required to complete a vulnerability assessment and Emergency Response Plan (ERP) under the Public Health Security and Bioterrorism Preparedness and Response Act (Bioterrorism Act, 2002). Nevertheless, utilities approach emergency planning and preparedness with varying levels of rigor, and some utilities have never needed to activate their emergency response systems. The MWRA Shaft 5A pipe break provides a window of opportunity to characterize and assess the effectiveness of a major utility's emergency response program based on its response to a significant emergency event. Prepared by the Association of Metropolitan Water Agencies, Water Research Foundation, WaterISAC; funded through a grant from the Alfred P. Sloan Foundation.
Publication 166—U.S. Postal Inspection Service Guide to Mail Center Security	http://about.usps.com/publications/pub166/welcome.htm	The U.S. Postal Inspection Service offers this guide to help mail center supervisors and co-workers keep mail center safe and secure. The guide provides general advice and recommends protective measures to help assess, prevent, and respond to three types of threats: mail theft; package bomb, or bomb threats; and chemical, biological, or radiological threats.

7.0 REFERENCES

Baron, G., & Philbin, J. (2019, April). Crisis communications in the age of Twitter. *Tactics*, 9.

Beaumont, C. (2019, January 16). New York plane crash: Twitter breaks the news, again. http://www.telegraph.co.uk/technology/twitter/4269765/New-York-plane-crash-Twitter-breaks-the-news-again.html

Bloomberg Businessweek. (2007). Responding to a crisis: What a business must do to handle a crisis in the public arena.

Covello, V., Minamyer, S., & Clayton, C. (2007). *Effective risk and crisis communication during water security emergencies*. U.S. Environmental Protection Agency.

4

Disaster Recovery

1.0 INTRODUCTION—WHAT IS DISASTER RECOVERY? 157
- 1.1 Reversing the Trend of Declining Community Resilience 159

2.0 PLANNING AND MANAGING SMALL-SCALE AND COMMUNITYWIDE DISASTER RECOVERY 159
- 2.1 Small-Scale Disaster Recovery 160
  - *2.1.1 Utility Integrated Contingency Plan Primary Objectives* 161
  - *2.1.2 Advantages of the Utility Integrated Contingency Plan Approach* 161
  - *2.1.3 Planning Characteristics of Utility's Integrated Contingency Plan* 161
- 2.2 Communitywide Recovery 162
- 2.3 The Importance of Recovery Preparation and Planning 162
  - *2.3.1 Establishing the Recovery Planning Team* 162
  - *2.3.2 Recovery Planning Process* 164
    - *2.3.2.1 Recovery Planning and Coordination With Local Plans* 165
      - *2.3.2.1.1 Public Works Plans and Policies* 165
      - *2.3.2.1.2 Emergency Management Plans and Policies* 166
      - *2.3.2.1.3 Economic Development Plans and Policies* 166
      - *2.3.2.1.4 Finance Plans and Policies* 166
      - *2.3.2.1.5 Other Local Plans and Policies* 166
    - *2.3.2.2 Recovery Planning—Staffing Needs* 167
    - *2.3.2.3 Recovery Planning—Legal Issues and Tasks* 167
    - *2.3.2.4 Recovery Planning—Finance Issues, Priorities, and Tasks* 168

2.3.2.5 *Documentation Requirements and Priorities* 169

2.3.3 *Planning for Critical Recovery Operations* 169

2.3.3.1 *Damage Assessment* 169

2.3.3.2 *Damage Assessment Memoranda of Understanding With Surrounding Utilities and Communities* 170

2.3.3.3 *Debris Management Planning and Multijurisdictional Memoranda of Understanding* 170

2.3.3.4 *State Department of Transportation Memoranda of Understanding* 170

2.3.4 *The Recovery Plan* 171

2.3.5 *Planning for Resource Needs* 171

2.3.5.1 *State and Local Recovery Resources* 171

2.3.5.2 *Local Recovery Funding* 172

2.3.5.3 *Few Disasters Are Supported by Presidential Declarations* 172

2.3.5.4 *Local Recovery Funding Options* 173

2.3.5.4.1 *Reserve Funds* 173

2.3.5.4.2 *Risk Abatement and Risk Mitigation* 173

2.3.5.4.3 *Risk Spreading* 173

2.3.5.4.4 *Risk Acceptance* 174

2.3.6 *Establishing Recovery Priorities: Conducting Damage Assessments* 174

2.3.7 *Why Plan for Recovery Damage Assessment?* 174

2.3.8 *Information Gathered by the Recovery Task Force and Others* 174

2.3.8.1 *Using Hazard Analysis Information to Plan for Recovery Implementation* 175

2.3.8.2 *Critical Infrastructure and Key Resources Risk Assessment* 175

2.3.8.3 *Determining Critical Infrastructure and Key Resources Recovery Priorities* 175

2.3.8.4 *Economic Effects* 176

2.3.8.5 *Community Needs* 176

2.3.8.6 *Public Health Concerns* 176

2.3.9 *Identifying Potential Recovery Projects* 176

2.3.10 *Prioritization of Recovery Projects* 177

2.3.10.1 *Selecting Recovery Projects to Implement* 177

2.3.10.2 *Implementation Criteria* 178

2.3.10.3 *Recovery Priorities: Providing Essential Services (Maintaining Services)* 178

2.3.10.4 *Recovery Priorities: Restoring Critical Infrastructure and Key Resources* 179
2.3.10.5 *Recovery Priorities: Protecting Public Safety and Health* 179
2.3.10.6 *Recovery Priorities: Protecting the Environment* 179
3.0 RECOVERY PLAN AND RECOVERY OPERATIONS PLAN ACTIVATION 180
3.1 Inspection of Damaged Infrastructure 180
3.2 Providing Support Services 181
3.3 Recovery Coordination and Communication 181
3.3.1 *Communication During Recovery* 182
3.3.2 *National Incident Management System* 183
3.3.3 *Intra-Agency Communication* 183
3.3.4 *Communication With the Public* 183
3.3.5 *Recovery Implementation* 184
3.4 Recovery Training 184
4.0 NATIONAL DISASTER RECOVERY POLICY 185
5.0 REFERENCES 187
6.0 SUGGESTED READINGS 190

1.0 INTRODUCTION—WHAT IS DISASTER RECOVERY?

In the United States in September 2011, recovery policy took on new status with formal posting of the National Disaster Recovery Framework by the Federal Emergency Management Agency (FEMA). The posting was formal acknowledgment that disaster recovery was a significant emergency management element on equal footing with preexisting prevention, protection, mitigation, and response policies (FEMA, 2011b).

> "We know how to achieve comprehensive emergency management—through preparedness, response, recovery and very importantly, prevention through mitigation."
>
> *James Lee Witt,*
> *FEMA Director nomination hearing,*
> *March 31, 1993* (U.S. Senate, 1993)

Those who are responsible for disaster recovery may suggest that it is understood and justified that recovery policy would be the last stand-alone national disaster policy to be developed, as recovery is the most long-lasting, complex, costly, and challenging phase of emergency management.

Even before FEMA adopted the National Disaster Recovery Framework (NDRF) in 2011, many local governments, utilities, and private and nonprofit partners were aggressively engaged in pre-disaster recovery planning. Communities throughout the United States have adopted NDRF policies (as draft policies and now as formal policy). Several examples are Florida counties of Hillsborough, Sarasota, Polk, Manatee, and Nassau; Panama City, Florida; Fairfax County, Virginia; and Beaufort County, South Carolina. It is important to note that community water and wastewater utilities are key participants and leaders in these community recovery plans, because of the critical need for community support, participation, and commitment to water and wastewater disaster recovery.

Predisaster recovery planning is based on the following policies that local governments and local private and nonprofit partners use to plan for communitywide needs in a post-disaster environment:

1. Establish an organizational and community-recovery planning framework that aligns with FEMA's National Disaster Recovery Framework (NDRF);
2. Development of recovery and operational planning that are consistent with the National Incident Management System (NIMS), and designates roles and responsibilities to coordinating and cooperating agencies and organizations;
3. Designation of recovery functions, pre-disaster objectives and strategies, and short-term recovery objectives and strategies (post-disaster) to enable community leaders to take action immediately following a disaster; and
4. Development of a comprehensive training agenda that includes NDRF policy, Incident Command System/NIMS interaction, recovery function policies, and operational control policies (documentation, spending procedures, and recovery ordinance requirements).

Details of the recovery planning and preparation process are discussed in the remainder of this chapter.

Disaster recovery planning can be focused on small-scale repairs and restoration, all the way to large-scale, long-term community reconstruction and redevelopment (Webber & Slaughter, 2010). Effective recovery plans and policies are "scalable" to meet the complexity, variety, and growing regionalization of disasters (Cutter, 2001). A disaster may strike quickly, leaving the need for recovery operations in its wake or it can be a prolonged event requiring recovery activities to begin while the response phase is still in full activation. Severe windstorms, fires, or floods are examples of disasters that can be ongoing, presenting recovery challenges during and after an event. A significant earthquake is an example of a disaster that strikes and is over quickly, leaving severe damage in its wake. Communitywide recovery is complex and often presents multiple disasters at once, also referred to as packages of disasters. For instance, a significant coastal hurricane includes high wind damage, storm surge effects, high volumes of debris, significantly damaged infrastructure, and multijurisdictional effects.

Communitywide recovery incorporates consideration of hazard mitigation throughout the recovery process. Successful communitywide recovery includes participation of affected communities to the extent possible. Recovery challenges vary significantly based on the magnitude of a disaster and resilience of the community and infrastructure. It may seem simplistic, but greater attention and focus on policies can reduce disaster effects and spending (Leonard & Howill, 2009; Phillips & Neal, 2007). The environment should be designed to allow the community and the individual to continue to function at an acceptable level and to mitigate any adverse consequences.

1.1 Reversing the Trend of Declining Community Resilience

People can increase their perceived and actual control by becoming actively involved in decision-making. "Learned-resourcefulness" refers to what a group can do to when stressful circumstances call for self-direction. Facilitating learned resourcefulness requires

- developing understanding of relevant functions and activities rather than focusing on hazard effects that may be difficult to comprehend;
- making resources available to support community-led development including infrastructure recovery, and
- developing ways for demands to be seen as challenges, that are worthy of investment and engagement, and meaningful.

2.0 PLANNING AND MANAGING SMALL-SCALE AND COMMUNITYWIDE DISASTER RECOVERY

Returning to normal (or near-normal) after a disaster does not just happen. Water resource recovery facilities are faced with threats, such as contamination, distribution system destruction, and equipment failure. Utilities are key communitywide recovery participants because of the critical role played in community recovery and the skills and training of staff. The effects of small-scale incidents are typically managed internally, whereas disasters are complex communitywide, multijurisdictional, and often regional disasters. In communitywide disasters, utilities become participants (partners) in

FIGURE 4.1 Infrastructure Resilience Properties (Adapted from Moore et al., 2016 with permission)

recovery. Recovery of this scale requires communitywide planning, priority setting, and cooperation (FEMA, 2011b).

2.1 Small-Scale Disaster Recovery

One of the most significant effects after a disaster can be damage to onsite wastewater system components. Tanks may be overwashed and contents discharged, or the tanks can be shifted, disconnected, and filled with sand or silt. When septic tanks or, more commonly, drain fields have eroded, there is often insufficient area remaining to repair the systems in accordance with local codes. Exposed tanks also can be a hazard to recovery workers or people returning to their properties. Although these are small-scale issues, they are critical to wastewater recovery and typically managed by the utility itself (Payne et al., 2007). The utility integrated contingency plan approach prepares a utility to plan for small-scale and communitywide recovery (U.S. Environmental Protection Agency [U.S. EPA], 2007). Regardless of the size and complexity of a disaster, geographic information support (GIS) staff capability has vastly improved recovery damage assessments, recovery planning, and recovery (Mohammed, 2008).

2.1.1 Utility Integrated Contingency Plan Primary Objectives

- To manage all response, resumption, recovery, and restoration activities;
- To protect employees and information assets until regular business operations are resumed;
- To ensure that a viable capability exists to respond to an incident;
- To support and communicate with employees, system administrators, security officers, and managers;
- To accomplish rapid and efficient resumption of time-sensitive business operations, technology, and functional support areas;
- To ensure regulatory requirements are satisfied;
- To exercise resumption and recovery expenditure and decision-making; and
- To streamline the reporting of resumption and recovery progress between recovery teams and management.

2.1.2 Advantages of the Utility Integrated Contingency Plan Approach

- Combines many plans and regulations into a single plan that meets all needs;
- Reduces errors by decision-makers regarding which of their plans is applicable to a particular emergency;
- Is a functionally friendly approach for use in varied emergency situations, while providing a mechanism to ensure compliance with multiple agency requirements; and
- Improves coordination among local response/recovery personnel; and local, state, and federal emergency response and recovery personnel.

2.1.3 Planning Characteristics of Utility's Integrated Contingency Plan

- Planning begins before disaster;
- Policies are scalable to manage small-scale and communitywide complex incidents;
- Prepares utilities to participate in the implementation of community-wide recovery planning and recovery; and
- Prepares water and wastewater utilities to
  - Simultaneously address multithreat scenarios facing communities;
  - Establish working relationships with economic development organizations to address threats to critical customers important to an entire community and the economic vitality and community tax and job base;

- Be effective leaders and participants in the implementation of communitywide planning and recovery implementation; and
- Provide assistance and support to other community recovery stakeholders.

Material found in Sections 2.1.1 through 2.1.3 came from U.S. EPA, 2007.

2.2 Communitywide Recovery

Effective communitywide recovery requires coordination among local governments, agencies, public and private utility providers, citizens, private sector, voluntary organizations, neighboring jurisdictions, and state and federal governments. The recovery process requires planning by more than local emergency management and disaster recovery agencies and public and private utilities. All community participants must work together seamlessly. Recovery plans and agreements (mutual aid, debris removal, damage assessment, etc.) must be in place well in advance of disasters. Water and wastewater utility disasters often have community effects that require community participation and support. Large-scale and smaller-scale recovery are well served by formal and informal cooperative relationships between communities and utilities (Haddow et al., 2008; Phillips & Neal, 2007).

2.3 The Importance of Recovery Preparation and Planning

Planning for recovery provides many gains for the community, including

- Organization, cooperation, partnerships and coordination;
- Incorporation of many points of view;
- Focus—a clear path for recovery;
- Community participation;
- An opportunity to improve community resilience through mitigation of community risks; and
- Community healing and a focus on the future (FEMA, 2005, 2011a).

2.3.1 Establishing the Recovery Planning Team

A local recovery task force (RTF) should not be too large, but should include public- and private-sector members. The RTF plays a significant role as a sounding board for new ideas and policies (FEMA, 2005). Membership in an RTF is a common way to establish the interagency cooperation needed in community and utility recovery. An RTF is often authorized by ordinance.

A recovery ordinance provides a framework for recovery planning and policies, including recovery task force management structure. Table 4.1 lists typical ordinance topics.

An RTF will participate in developing policies and procedures for community recovery and redevelopment. An RTF will help identify mitigation opportunities and review disaster status reports. The team may also make recommendations to modify ordinances and policies (zoning, permitting, etc.) to enhance recovery. Utilities should include their stakeholders as participants in the planning process; community planning teams should include supervisors or functional directors from utilities and local government(s), including public works, public facilities, sanitation, and other groups, as needed. Stakeholders and partners may also include

- Emergency management representatives or liaisons,
- Other water/wastewater agencies (local and regional),
- Private utilities and engineers (local and regional),
- Department of health representatives,
- Corresponding state agencies,
- Historic preservation officers and environmental liaisons, and
- Financial and procurement representatives (FEMA, 2012a, 2012b).

Table 4.2 includes a list of disaster-recovery task force participants.

TABLE 4.1 Disaster Recovery Ordinance Content

Typical topics and policies
Recovery task force management and structure
Recovery planning policies
Recovery functions (activities)
Operational policy
Redevelopment authority
Power to recommend county ordinance and policy revisions needed to support recovery
Authority to establish regional participation in local recovery planning
Creation of special committees
Interfacing with state and federal representatives

TABLE 4.2 Disaster Recovery Task Force Participants

Chair of county/city council, county/city administrator, deputy county/city administrator
County/city attorney, recovery director, and deputy director
Emergency management director and deputy director
Private sector: chambers of commerce, downtown association, critical community employers, lenders, real estate
County/city recovery function leaders
Economic development organizations
School district, military facilities, state agencies, wastewater, water, gas, electricity utilities
Community groups, Red Cross, Salvation Army, regional transportation, religious groups/nonprofits

2.3.2 Recovery Planning Process

Recovery planning is a community planning process that includes the public and private sectors. Many communities create an RTF with broad community representation. Community involvement improves the quality of recovery projects and helps overcome loss of pride and community spirit. Too many communities ignore the value of community pride and confidence. It is necessary to manage the planning process and to keep the public informed of issues that may affect the recovery schedule. Excessive schedule delays can weaken community confidence and drive. The planning team must ensure that the public understands that the plan will be revised as recovery unfolds. It is necessary to facilitate realistic thinking about the time it will take to implement the recovery plan. Recovery may take years if the disaster is a large one (FEMA, 2005). It is up to the planning team to assign recovery roles and responsibilities. These decisions provide a framework to accomplish the administrative duties needed in recovery.

Recovery planning administrative duties include

- Recovery plan coordination with other local plans,
- Staffing needs and roles,
- Legal issues and tasks,
- Financial priorities and tasks, and
- Documentation requirements and priorities (FEMA, 2012a, 2012b).

Public involvement in recovery planning is an effective recovery policy:

- Involvement will more likely lead to inclusive planning with community support;
- Town hall meetings, surveys, online forums, and social media are effective tools for public recovery and mitigation projects input;
- Communicate instructions for community participation in these recovery activities, and why these activities are important to the process;
- Public involvement can reach beyond traditional jurisdictional boundaries to establish the basis for regional collaboration;
- Effective communication and coordination go hand in hand;
- Public information efforts must be coordinated with other agencies, organizations, and jurisdictions to ensure that clear and consistent messages are conveyed; and
- Public involvement in the recovery process generates buy-in and leads to success (FEMA, 2012a; Smith & Wenger, 2006).

2.3.2.1 *Recovery Planning and Coordination With Local Plans*

Coordinate city/county recovery plans with other local plans (and regional plans, if necessary) in the following categories:

- Public works,
- Emergency management,
- Economic development, and
- Finance.

2.3.2.1.1 Public Works Plans and Policies

Public works plans and activities are guided by the following plans and ordinances:

- Floodplain management plan,
- Stormwater management plan,
- Debris management plan,
- Capital improvements plan,
- Subdivision ordinances,
- Land development code, and
- Building codes.

2.3.2.1.2 Emergency Management Plans and Policies

- Community disaster recovery and reconstruction plan,
- Comprehensive emergency management plan or emergency operations plan,
- Continuity of operations plan,
- Continuity of government plan,
- Local or regional evacuation plan,
- Natural hazard mitigation plan,
- Flood response plan,
- Temporary disaster housing plan,
- National Flood Insurance Program (NFIP) plan,
- NFIP community rating system policy plan, and
- Special needs populations plan.

2.3.2.1.3 Economic Development Plans and Policies

- Community and regional economic development plan;
- Business recruitment and retention plan;
- Local development corporation business development plan;
- U.S. Small Business Administration (SBA) loan program (packaging SBA loans);
- Downtown revitalization plan;
- Business expansion financing program;
- Business and industry equipment financing program; and
- Business marketing plans (of local governments, chambers of commerce, and regional development organizations).

2.3.2.1.4 Finance Plans and Policies

Coordinate recovery plans with plans of the local government's budget agency or department; finance staff members are key recovery participants.

2.3.2.1.5 Other Local Plans and Policies

- Regional, county, and municipal comprehensive plan;
- Open space management plan;
- Natural resource protection plan;
- Comprehensive land use plan;
- Historic preservation plan;

- Zoning ordinances; and
- Fire code.

Information in this entire section (2.3.2.1 through 2.3.2.1.5) comes from FEMA, 2012a, 2012b.

2.3.2.2 Recovery Planning—Staffing Needs

Identify a planning coordinator who will be responsible for assigning roles and responsibilities associated with planning. Tasks associated with staffing considerations include

- Preparing the plan,
- Reviewing and approving the plan (developing processes for review and approval),
- Identifying staffing needs (to meet needs as they are identified),
- Identifying training needs,
- Coordinating with state and federal agencies,
- Initiating mutual aid agreements and contracts,
- Coordinating with emergency management and other agencies, and
- Determining appropriate use and management of volunteers (FEMA, 2005).

2.3.2.3 Recovery Planning—Legal Issues and Tasks

There are significant legal considerations related to disaster recovery. All local professionals, including utility professionals, should seek legal advice in the development of recovery plans, procedures, authorities, and policies. Tasks associated with legal considerations may include

- Reviewing ordinances that address and reflect recovery or other emergency management issues, such as building codes and permitting;
- Reviewing contracts and agreements, such as debris disposal contracts;
- Developing and monitoring emergency procurement codes;
- Assisting with document preparation, such as interagency agreements;
- Considering litigation and claims; and
- Documenting the procurement process.

Keep in mind that emergency procurement codes may allow purchasing items differently in an emergency than under typical circumstances. Proposal review and selection periods are likely to be shortened and credit limits may be raised. Research local procurement codes well before disaster strikes.

2.3.2.4 Recovery Planning—Finance Issues, Priorities, and Tasks

Before a disaster, a community and/or utility should consider whether it is possible to finance recovery or a portion of recovery costs. Federal financial resources tend to be narrowly focused, adding to the importance of local financing capacity-building (Smith, 2010). Funding mechanisms should include local resources, and federal, state, and nonprofit sources. Personnel should be assigned to carry out and coordinate financial issues that may arise during recovery planning and implementation. Tasks associated with financial considerations may include

- Estimating the cost of recovery activity and available local funding;
- Considering effect on local tax and revenue structures;
- Understanding potential reimbursable actions (and policies) in federal programs, such as those associated with the Stafford Act (see Suggested Readings section);
- Researching existing practices regarding cost-sharing arrangements with state and federal agencies, and pre-establishing cost-sharing arrangements as needed;
- Exploring other potential sources of revenue; and
- Monitoring emergency procurement (FEMA, 2012a).

Communities need to be smart in preparation for securing support and interest from agencies and organizations willing to invest in a project, a community, and a process. Successful communities approach this task as it they were seeking investors and public funds for their favorite community project. The community needs to put its best case forward and plan how it will seek support. The Federal Emergency Management Agency suggests establishing partnerships with the various state, federal, and not-for-profit agencies. Communities are more successful if they have developed relationships with potential partnership (and funding) agencies. This approach allows more time to become familiar with funding policies and limitations. It also allows the community to communicate and sell the project and the scope of redevelopment to the partnering agency. The goal of this approach is to lay the foundation for long-term relationships with investors, public agencies, and lenders (FEMA, 2005).

2.3.2.5 Documentation Requirements and Priorities

Proper documentation is vital to local reimbursement (from other sources) and effective recovery management. An important part of establishing documentation procedures is pre-identifying the agencies and individuals who will be responsible for documentation during a local event.

Consider documentation needs related to accountability and planning, such as

- Adherence to state and federal requirements,
- Project worksheet requirements,
- Progress reporting,
- Contract procedures for goods or services,
- Expenditures for recovery activity,
- All other agreements,
- Potential litigation and audit policies, and
- Electronic records and files.

When documenting, make multiple copies and keep files for local records. It is also best to have digital and/or hard copy backup files in a separate location. Information in this section was taken from FEMA, 2012b.

2.3.3 Planning for Critical Recovery Operations

Communities and utilities can do many things to be better prepared for successful recovery. Several policy areas are listed below (IEMC-E930, 2010a).

2.3.3.1 Damage Assessment

Damage assessment takes place in response and recovery; the scope and purposes are different.

Damage assessment is important to recovery preparation and implementation. Damage assessment begins early in the response phase, but does not end there. Assessment is also a key recovery function. The focus of assessment changes at several critical points in the disaster management cycle. As response begins, assessment data are used to establish emergency priorities and as the basis for seeking a presidential disaster declaration. These data also are shared with those entities that are responsible for recovery planning. Damage data is the lifeblood of effective disaster decision-making, including recovery. In fact, sound damage data are so important to recovery that utilities and communities (as a whole) need to consider developing communitywide damage assessment agreements and policies to manage and improve the recovery damage assessment phase of emergency management.

Recovery damage assessments are focused on detailed reports. The reporting will provide engineering data used in the project evaluation process. Reports will focus on costs and feasibility.

2.3.3.2 Damage Assessment Memoranda of Understanding With Surrounding Utilities and Communities

Disasters that have regional effects often expand the time required for recovery planning preparation. Many communities and utilities develop mutual aid agreements to help each other to collect affect data. Data gathering agreements may cover a region, county, municipalities, and special zones or areas, such as coastal areas that include several counties and municipalities (Cutter et al., 2000; Heinz, 2000). Often agreements will include the assignment of one locality's damage assessment teams to another community that has greater need. An effective feature of many successful countywide and/or regional damage assessment programs is a common digital data collection system. Often one municipality agrees to perform management information system (MIS)/GIS functions for everyone. Disaster declaration reports then can be developed for regions, or for one community (Mohammed, 2008).

2.3.3.3 Debris Management Planning and Multijurisdictional Memoranda of Understanding

Effective and timely debris management is a critical recovery factor. Although communities and utilities may have individual debris management vendor contracts, regional and communitywide debris site planning and utilization avoid useless overlapping and jurisdictional competition for sites. Using a debris management memorandum of understanding (MOU) for identifying and managing sites frees communities to focus on other community needs at a time when there are many priorities that need attention.

2.3.3.4 State Department of Transportation Memoranda of Understanding

Building on agreements with several municipalities for debris removal, a state road debris removal agreement with the state department of transportation will enable local governments or entire regions to remove debris from state roads faster, improving response and recovery speed and public safety. Often debris management MOUs have other features, such as

1. Joint field and tabletop training;
2. Staff sharing provisions; and
3. Unified policies for handling documentation required for reimbursement by federal, state, and nonprofit organizations.

2.3.4 The Recovery Plan

A recovery plan is a tool to guide communitywide and utility recovery processes. Successful recovery requires action and timely accomplishment. A recovery plan is a living document that will be amended many times as the process moves forward. Policy changes may also be needed if unanticipated issues are confronted. During the implementation of a recovery plan, projects will be planned, designed, funded, approved, and completed (IEMC-E930, 2010a).

The needed resources and funding must be reevaluated project by project. Development and implementation of a recovery plan establishes the basic recovery footprint for community and utility recovery (Smith, 2004).

2.3.5 Planning for Resource Needs

In many instances early in a significant event, without adequate resource planning, local governments and utilities are likely to be overwhelmed. Managers will need to plan and exercise a high level of creativity to maintain adequate resources to react to the demands of a community disaster. Unfortunately many community utilities, businesses, and other organizations do not prepare to meet these challenges and have unrealistic perceptions of the availability of federal and state disaster resources (Schwab et al., 1998).

2.3.5.1 State and Local Recovery Resources

State recovery funding varies from state to state. State financial assistance often follows federal programs, often providing a portion of matching funds for federal and nonprofit programs.

States often divert state resources to areas affected by significant disasters. An example might be a state infrastructure bank that is created as an economic development tool to assist localities in business recruitment and business expansion by providing infrastructure funding. However, in a disaster the bank may divert funds to infrastructure rebuilding in disaster areas. States also provide nonfinancial assistance. State departments often assign staff to provide assistance to individuals, families, businesses, and local governments. In some cases, local recovery assignments come from state recovery plans and policies (IEMC-E930, 2010b).

2.3.5.2 Local Recovery Funding

Effort should be made to leverage and to plan to effectively use local resources as much as possible. Even communities and utilities with limited resources can effectively repair infrastructure by adopting local policies, such as those below. When additional resources are needed, they may be gained by

- Redirecting existing in-house resources (from other uses),
- Mutual aid/MOUs for assistance (offsetting labor costs), and
- Contracts with private contractors (outsourcing) (IEMC-E930, 2010a).

Water/Wastewater Agency Response Network (WARN) mutual aid agreements offer communities an opportunity to secure assistance from neighboring communities and utilities at little or no cost. A WARN agreement does not require a presidential declaration for activation. Communities often consider WARN agreements as local recovery tools. Before large-scale events, communities and utilities need to procure or initiate contingent contracts for purchasing goods and services. Planning for a legally defensible emergency procurement process is a necessity. Establishing agreements, contracts, and emergency procurement procedures before an event will facilitate a more accountable and effective recovery and timely reimbursement from state, federal, and nonprofit recovery programs (U.S. EPA, 2007, 2008).

2.3.5.3 Few Disasters Are Supported by Presidential Declarations

Disaster events often result in damages that do not meet federal disaster declaration thresholds. Over the last decade, competition for declarations has also increased. Without declarations, federal recovery funds are not available. A small percentage of disasters are supported by a presidential declaration (Schwab et al., 1998). Without a presidential declaration, requests may be made to federal agencies under their own authority from existing or emergency programs, such as the SBA or the U.S. Department of Agriculture. A governor's request for assistance from federal agencies (under their own authority) from existing or emergency programs is transmitted directly to the agency involved. The specific assistance, if awarded, is then provided directly from the federal agency. In most cases, state and local governments, insurance companies, and citizens assume the cost of recovery. Local funding preparation needs to be in place and a high priority in community and utility recovery planning (Beaufort County, 2012).

2.3.5.4 *Local Recovery Funding Options*

What are the primary local funding options (Beaufort County, 2011, 2012)?

- Reserve funds,
- Lines of credit,
- Risk abatement and/or risk mitigation policies,
- Risk spreading policies, and
- Risk acceptance.

2.3.5.4.1 Reserve Funds

Many types of organizations create reserve funds to be prepared for expenses that are not planned expenses. These funds are accumulated from many sources, most commonly from a set percentage of revenue diverted to a fund dedicated to disaster spending. Reserve funds can also be used to leverage recovery resources by matching state, federal, and nonprofit grants.

2.3.5.4.2 Risk Abatement and Risk Mitigation

The goal of risk abatement and risk mitigation is to minimize risk to a level that is compatible with operating standards of a local government or utility. The risks are not eliminated, but the severity of the loss is reduced to a level that can be handled. Common abatement or mitigation examples are retrofitting existing structures and adopting new building construction standards to withstand higher wind and/or higher storm surges (May, 2008).

2.3.5.4.3 Risk Spreading

Another term for risk spreading is risk transfer. Local governments can transfer risk by purchasing insurance coverage to cover portions of disaster costs. The cost of doing business approach is another form of risk spreading that builds projected disaster costs into revenue-generating policies of an entire organization. Local governments often approach disaster funding from this perspective, however the potential need is understated. Local governments often do not dedicate these funds to disaster damage only. Communities often use funds from these sources to fund local projects that are not emergencies (May, 2008).

2.3.5.4.4 Risk Acceptance

Organizations often accept risk out of a sense of resignation. Often, this passive approach leads communities and organizations to ignore alternatives available to them that may reduce risk exposure. This attitude comes from the perspective that not all risk can be avoided, transferred, or abated, so why waste time and money. Risk acceptance based on effective risk assessment is, however, an acceptable approach (May, 2008).

2.3.6 Establishing Recovery Priorities: Conducting Damage Assessments

The Federal Emergency Management Agency suggests that communities and utilities answer two questions before beginning the damage assessment process:

- What process and standards will be used for assessment, documentation, and reporting?
- What critical facilities and key resources were affected, identified by hazard mitigation plans and hazards inventory and vulnerability analysis (HIVA) (FEMA, 2011a)?

Immediately following an event, one of the most significant challenges for communities and utilities (other than human needs) is detecting and isolating damages to critical infrastructure and facilities (Drabek & Evans, 2010). Assessment is a response and recovery function. How do communities and utilities prepare for this important activity? Several leading issues are deciding who does what, what information is needed, and who should receive the information.

2.3.7 Why Plan for Recovery Damage Assessment?

Planning for damage assessment will help to ensure that the process goes smoothly when a disaster strikes. It will also assist in "institutionalizing" damage assessment within organizations. The Federal Emergency Management Agency suggests failure to plan for damage detection often leads to failed presidential disaster declaration efforts. Failure is based on poor decision-making, wrong priorities, and undetected threats (FEMA, 2012a).

2.3.8 Information Gathered by the Recovery Task Force and Others

In addition to collecting damage data to plan effectively, the RTF requires information to set priorities, assign responsibilities, and allocate resources.

- Established pre-event priorities,
- Communitywide recovery plan(s) (if they exist),

- Community demographics (to anticipate needs of affected people/areas),
- Availability of resources (public and private sector),
- Inventory and availability of goods and supplies,
- Jurisdictional responsibilities of participants,
- Risk and vulnerability assessments (if they exist), and
- Pre-established services contracts (Beaufort County, 2012; FEMA, 2012b; IEMC-E930, 2010a).

2.3.8.1 Using Hazard Analysis Information to Plan for Recovery Implementation

Risk and vulnerability assessments include information about how often each hazard is likely to occur, areas likely to be affected, and severity of the effect. Although every community is different, some basic steps need to be performed to effectively assess risk and vulnerabilities of communities and utilities (FEMA, 2008a, 2009).

2.3.8.2 Critical Infrastructure and Key Resources Risk Assessment

Risk and vulnerability assessments identify the effect of an event on water and wastewater systems, as well as other critical infrastructure (and other community key resources). Critical infrastructure and key resources are necessary for the health and welfare of the population of a community. Critical infrastructure recovery often falls under the scope of water/wastewater utilities, which means they will lead many recovery projects. Restoration of essential services and repairs and rebuilding key assets are high priorities for community recovery. These components include public safety services, health care, utilities, transportation systems, and lifelines. In addition, critical infrastructure and key resources include facilities that, if affected by a hazard event, could result in high losses and/or release of hazardous materials, which would further result in significant environmental effects if not addressed in the assessment process (FEMA, 2008a, 2008b).

2.3.8.3 Determining Critical Infrastructure and Key Resources Recovery Priorities

Consideration should be given to all critical infrastructure and key resources identified in the most recent version of the local hazard mitigation plan or the community HIVA. Critical infrastructure and key resources include roads, bridges, utilities, and other community assets. Also consider pre-event utility priorities and needs, as established in risk and vulnerability assessments. Keep in mind that damage assessment information may identify new priorities. Additionally, consider planning criteria specifically related to utility recovery activities, such as utility restoration and water/wastewater treatment (FEMA, 2011a).

When establishing recovery priorities to include in a recovery plan, consider the typical prioritization of

- Critical infrastructure and key resources,
- Economic effects (and needs),
- Community needs, and
- Public health concerns (FEMA, 2011a).

2.3.8.4 Economic Effects

Significant disasters may have multiple of economic effects on a community. For example, when a storm disrupts and destroys water system capacities, restoring water system operations are essential for the economic recovery of businesses and the community. The economic effects of a communitywide disaster may take years to address, compounding economic effects that are based on job and tax base losses (FEMA, 2005, 2011b).

2.3.8.5 Community Needs

Those responsible for identifying community needs must build that information on basic community demographic information, such as the number of households, income levels, and the like. The number of households helps to anticipate temporary housing needs. Community needs come in many forms. Utilities and other organizations may be called on to support things, such as public shelters, with installing generators or distributing food and water (IEMC-E930, 2010a).

2.3.8.6 Public Health Concerns

Public health concerns are a significant consideration. A recovery plan should cover keeping essential services operational. For example, the interruption of services, such as routine household solid waste collection or a water resource recovery facility (WRRF), could all have negative effects on public health. By planning for these public health concerns, the community will be better able to recover and to reduce potential for secondary disaster effects (FEMA, 2012b).

2.3.9 Identifying Potential Recovery Projects

Damages and their location, long-term effects, housing and infrastructure damages, and community economic needs are some of the core factors to consider. In the confusion after a significant disaster, simplifying the task is necessary. An effective tactic is to organize damages into short- and long-term recovery objectives. Short-term recovery projects are typically those that allow citizens and businesses to reinhabit their homes, neighborhoods, and business areas. Before these things can happen, roads need to be open and passable, utilities operational, debris removed from roadways, and buildings classed as habitable. Long-term recovery projects often involve activities and actions taken to reduce future vulnerability and to return the community to predisaster conditions, or as near predisaster conditions as possible (Beaufort County, 2011; FEMA, 2012a).

2.3.10 Prioritization of Recovery Projects

Projects that can be accomplished relatively quickly with solid community support and available financial resources should begin as soon as possible. Completion of these projects will foster and strengthen community support and political endorsement of the recovery plan, recovery policies, and recovery projects. The end result of such swift and positive action increases the possibility of a comprehensive recovery. Projects that have the most significant effect on communitywide recovery should have highest priority. Those projects with the highest recovery value to local residents and community leaders should be pursued and completed as soon as possible. Water and wastewater recovery projects are often at the top of the recovery projects lists. Restoration of essential services, repair and reconstruction of critical infrastructure, and provision of services must be accomplished or communitywide recovery may be significantly delayed, or come to an unexpected halt. For example, choosing between two important recovery projects, such as the reconstruction of a community center or the strengthening of a WRRF, is a difficult decision because both are important to the community. However, the essential service provided by the WRRF makes it a higher priority. Keep in mind that although predefined recovery priorities are critical to the recovery effort, the recovery plan has to remain flexible. It should be considered a guide or framework and may evolve as the recovery process continues.

2.3.10.1 Selecting Recovery Projects to Implement

Identifying recovery projects typically results in a comprehensive projects list, and it is unlikely that there will be enough funding for all identified recovery projects. The resulting competition for funding must be managed and structured professionally to ensure a unified recovery effort. Selection should be based on priorities set with criteria that include life safety issues, restoration of services critical to the community, and reconstruction of critical infrastructure important to the local economy and the community's quality of life. Ultimately, the evaluation process must consider

- Newly established priorities,
- Infrastructure services,
- Overall community recovery effort or defined goals, and
- Cost and available financial resources (FEMA, 2012a; IEMC-E930, 2010a).

2.3.10.2 Implementation Criteria

Project review criteria should be established to review and rate each potential project. The system should have a focus on the needs of the community, project feasibility, and sustainability. Review criteria should include the following categories:

- Quality of life,
- Economic effect,
- Social equity,
- Environmental concerns,
- Hazard mitigation value,
- Community support,
- Relation to other projects,
- Time to complete, and
- Funding options.

Examples of recovery projects may include utility upgrades, expanded facilities, and relocation of highways or roads (FEMA, 2012a).

2.3.10.3 Recovery Priorities: Providing Essential Services (Maintaining Services)

Following a disaster, a utility should continue to provide day-to-day services if possible; this requires prior planning and backup systems and staffing. Diligent efforts should be made to return the community to normal; it will aid communitywide recovery and possibly prevent secondary disasters. To ignore certain day-to-day operations, such as contamination and sanitation issues, may create additional and unnecessary issues that require involvement of other government agencies (e.g., public health) and can inadvertently slow community recovery. The restoration of regular local government and utility services are key aspects of recovery. Damaged lifelines, such as utilities, roads and bridges, or public facilities should be identified and prioritized during the damage assessment process (Beaufort County, 2011; FEMA, 2012a).

2.3.10.4 Recovery Priorities: Restoring Critical Infrastructure and Key Resources

Critical infrastructure and key resources should be identified in vulnerability assessments. Consideration of the effect of community exposure to hazards will help set pre-event priorities for restoration or reconstruction. After a disaster, with current event damage assessments in hand, new priorities for repair and reconstruction may be identified. Critical infrastructure and key resources that may be included are

- Water and wastewater treatment, distribution, and collection;
- Surface water systems;
- Highways, roads, and bridges (transportation systems);
- Public transit, railroads, and airports;
- Communications facilities;
- Public buildings and facilities; and
- Solid waste facilities.

2.3.10.5 Recovery Priorities: Protecting Public Safety and Health

Protection of life is the top priority for all local governmental agencies, including utilities. There are recovery public safety and public health issues that utilities may be assigned to

- Hazardous materials incidents,
- Animal and vector control, and
- Secondary threats or hazards.

2.3.10.6 Recovery Priorities: Protecting the Environment

Environmental concerns may exist that require water/wastewater utilities to act as the lead agency in areas, such as

- Addressing contamination of surface water or drinking water;
- Disposal of household hazardous materials;
- Clearing and cleaning canals and other storm drainage systems;

- Debris management;
- Revegetation;
- Coastal management, such as beach renourishment; and
- Wetlands revitalization (Beaufort County, 2011; FEMA, 2012a).

3.0 RECOVERY PLAN AND RECOVERY OPERATIONS PLAN ACTIVATION

Plan activation occurs when a local government or utility determines that response is ending. Complex large-scale recovery transition from response to recovery will likely take place over time and be guided by a "handover checklist" (and other recovery operations plan polices) (Beaufort County, 2012).

Operational functions include assigning recovery roles, activating recovery functions (and extending key emergency support functions); coordinating emergency operations center operations; recovery planning; volunteers; and programs and staff with state, federal, and nonprofit recovery contacts.

3.1 Inspection of Damaged Infrastructure

In community recovery, water and wastewater utilities may be key participants. Inspection services involve the determination of whether repair or reconstruction is allowed for damaged infrastructure, buildings, and facilities. Inspection services also support communitywide effort, providing assistance to other agencies and supporting all recovery management activity. During large-scale recovery, inspection services may provide technical support to other agencies. The basic issues or functions of inspection services may include

- Damage assessment;
- Inspection priorities;
- In-house versus contracting for services;
- Use of outside resources, such as U.S. EPA's Multi-Hazard Mitigation or similar programs;
- Issuance of building permits;
- Post-disaster permitting procedures;
- Determination of habitability;
- Policy or procedure development;
- Local management of contractors;
- Code or ordinance review; and

- Training and exercise participation (Beaufort County, 2011; FEMA, 2012a).

Inspection of damages to public infrastructure, homes, businesses, industry, or other private structures are critical recovery tasks. These services may include

- Establishment of inspection request procedures;
- Inspections for life safety and structural integrity, code compliance, and reconstruction or rebuilding activities; and
- Establishment of a building tagging system.

Inspection services should ensure that the public is informed of their procedures for inspection, permitting, and technical assistance. They should work within the established public information system. Inspection services should also secure documentation of all activity as called for in the community's established emergency response and recovery documentation process.

3.2 Providing Support Services

In addition to the recovery efforts previously described, a water/waste water utility may provide support to other local agencies. Possible support services may include

- Support of community-wide recovery,
- Technical and engineering services,
- Temporary repairs,
- Cost estimates,
- Status updates,
- Ordinance and codes evaluation (to support recovery),
- Dissemination of information to the public,
- Coordination with state or federal agencies, and
- Evaluation of lessons learned,

3.3 Recovery Coordination and Communication

After a large-scale, and some categories of smaller-scale disasters, many different individual agencies and organizations (and often multiple jurisdictions) must work together to restore affected areas to as near predisaster condition as possible, or to an improved condition. Coordination and communication are keys to successful disaster recovery (Crandall & Spillman, 2010).

Public and private water/wastewater utilities are critical participants in the coordination of local recovery and, in particular, recovery projects. Utilities need to be prepared to coordinate water/wastewater recovery with communitywide recovery. It is prudent to assign the coordination responsibility to one or more persons within the utility. To effectively coordinate recovery of water/wastewater utilities with community recovery, consider using the following methods:

- Because some projects may require public and private financial support, public–private partnerships should be considered for coordination of projects;
- Assign dedicated personnel and/or liaisons to the community recovery function;
- Ensure that the utility actively participates as a partner in community recovery, which will include notices for meetings, review of policies and issues, and so forth;
- Proactively interacting with critical recovery partners to formally establish the liaison role for water/wastewater utilities;
- Personally inquire about and attend all recovery meetings, even at times when water/wastewater is not typically included. Do not be reluctant to insert water/wastewater utilities into the overall recovery process; and
- Ensure that existing MOUs and mutual aid agreements are current and continue to meet identified needs (FEMA, 2012b).

3.3.1 Communication During Recovery

Communication focused on recovery work, such as schedules, timelines, and warnings, is critical. Lack of this type of communication is likely to delay recovery and frustrate other agencies and the public. Public and private water/wastewater utility public information officer (PIO), in cooperation with the local agency PIO must keep local, state, and federal agencies and the public informed. Communication following an event or disaster is important to recovery. Depending on the nature of the event, communication and public information related to the event will potentially need to continue for months or years. The communications mission is to deliver accurate information in a timely manner, with an emphasis on reconstruction and saving lives and property. The community should use all communication strategies defined by the National Incident Management System (Crandall & Spillman, 2010; FEMA, 2012a).

3.3.2 National Incident Management System

The NIMS provides a systematic, proactive approach to guide departments and agencies at all levels of government, nongovernmental organizations, and the private sector. National Incident Management System policies work hand in hand with the NDRF and the National Response Framework. The NIMS provides the template for the management of incidents, whereas the NDRF provides the structure and mechanisms for coordinated policy for recovery incident management (FEMA, 2012b).

National Incident Management System training is a significant recovery element, as communities work to expand local, state, and national coordination.

3.3.3 Intra-Agency Communication

Communications with other agencies, such as local public works entities and state and federal liaisons, should, at a minimum, include

- Updates to current disaster information,
- Damage assessment updates,
- Strategies or plans for recovery,
- Specific resource needs,
- Financial effects,
- Environmental concerns, and
- Changes to service levels.

Because of its urgent nature, water/wastewater information is often the most important information an emergency manager or PIO may receive during recovery, and water/wastewater utilities should not wait to be asked for updates. Utilities should automatically update the PIO/emergency manager when there is information that needs to be reported (FEMA, 2012a).

3.3.4 Communication With the Public

Throughout the recovery process, the public should continue to be informed about activities and plans for the community. Information that should be conveyed to the public during recovery includes warnings or potential life safety issues; recovery plans that exist within the community; and how the community has been or is expected to be affected because of changes in services or infrastructure reconstruction. The public should also be aware of where and how to find additional information about the recovery process and programs, such as individual assistance programs. Some of the outlets for dissemination of information include news releases, public service announcements, social media, and recovery websites. During recovery, it is critical to have a water/wastewater utility liaison appointed to the public information recovery function to ensure that the public is informed on a timely basis (Crandall & Spillman, 2010; FEMA, 2012a).

3.3.5 Recovery Implementation

After a disaster, there will likely be many competing priorities. It is important that water/wastewater utilities and the community as a whole are able to choose the recovery actions that will best meet community and utility recovery goals, while working within budget, time constraints, and changing conditions. The plan needs to be updated when priorities change or new information becomes available. Maintenance of the recovery plan also includes review of basic components, including administration, planning, plan development and implementation, resource management, communication, and public information. Documentation of project implementation is essential to successful recovery and acknowledging lessons learned. It is also necessary to ensure accountability of expenditure of public funds. Because the recovery process is often accomplished in years rather than months, there are other considerations in maintaining an active and continuing recovery plan. Some examples are

- New priorities,
- Revised policies and procedures,
- Analysis criteria (based on changing conditions),
- Process or plan development,
- Training and exercises,
- Roles and responsibilities, and
- Funding availability (Beaufort County, 2012; FEMA, 2012b).

3.4 Recovery Training

Without recovery training and feedback, a recovery plan is little more than a report. Implementation is the key to community and water/wastewater utility recovery. If projects related to essential community water/wastewater services and critical infrastructure are not completed, recovery halts. U.S. EPA water/wastewater tabletop and NIMS training policies illustrate the linkage between recovery training and recovery success. Without training, the complexity of recovery works against recovery effectiveness. The NIMS crisis training focuses on technical knowledge, policy awareness, and effective communication. Crisis communication training includes media training for interaction with newspapers; magazines; and local, regional, and national television. Training builds resilience. Repetition is the basis of learning (Crandall & Spillman, 2010; FEMA, 2012; U.S. EPA, 2008).

Tabletop exercises typically are aimed at facilitating understanding of concepts, identifying strengths and shortcomings, and changing the approach to a particular situation. Participants are encouraged to discuss issues in depth and to arrive at decisions through slow-paced problem solving, rather than the rapid, spontaneous decision-making that occurs under actual or simulated emergency conditions. Practicing and exercising a WARN mutual aid agreement helps member utilities understand how the agreement may be implemented. By practicing the activation of an agreement and operational plans, a utility can respond more effectively and efficiently during an actual incident. Additionally, tabletop exercises facilitate opportunities to

- Build relationships among participants;
- Gain working knowledge related to roles and responsibilities in the WARN process, including the involvement of utilities, the state drinking water primacy agency, state wastewater permitting authority, state and local emergency management and response agencies, U.S. EPA (as appropriate), and supporting professional organizations;
- Test agreement provisions and operations policy and make changes, if needed;
- Explore effectiveness mobilization and demobilization policies; and
- Test agreement activation and policies for requesting and matching resources.

Information in this section comes from U.S. EPA, 2008.

4.0 NATIONAL DISASTER RECOVERY POLICY

National disaster recovery policy has taken significant steps forward. In September 2011, the NDRF was formally posted. A second document, the NDRF Interagency Operational Plan, recommends policies and roles of federal agencies and departments necessary to carry out NDRF objectives.

States, local governments, utilities, businesses, and other organizations that are recovery stakeholders need to be aware of national disaster recovery policies and programs. The NDRF and the NDRF Interagency Operational Plan define

- Core recovery principles;
- A coordinating structure that facilitates communication and collaboration among all stakeholders and guidance for pre- and post-disaster recovery planning;
- Roles and responsibilities of recovery coordinators and other stakeholders;
- The overall process by which communities can capitalize on opportunities to rebuild stronger, smarter, and safer; and
- Policy related to recovery support functions (RSFs).

The framework is a conceptual guide to promote coordination and recovery planning at all levels of government. The NDRF establishes coordination structures, and defines leadership roles and responsibilities. The NDRF provides a flexible structure that enables disaster recovery managers to operate in a unified and collaborative manner. It also focuses on how best to restore, redevelop, and revitalize the health, social, economic, natural, and environmental fabric of communities.

The NDRF is consistent with the vision set forth in the Presidential Policy Directive (PPD)-8, which directed FEMA to work with interagency partners to publish the recovery framework. It is the first policy framework published under the PPD reflecting core recovery capabilities by supporting operational plans as an integral element of the National Preparedness System. It is a first step toward the PPD-8 objective to achieve a shared understanding and a common, integrated perspective across all mission areas—prevention, protection, mitigation, response, and recovery—to achieve unity of effort and most effectively use the nation's limited resources. The material for this section came from FEMA (2011b). The following six NDRF RSFs and lead agencies are:

- RSF 1—Community Planning and Capacity Building—FEMA,
- RSF 2—Economic—U.S. Department of Commerce,
- RSF 3—Health and Social Services—U.S. Department of Health and Human Services,
- RSF 4—Housing—U.S. Department of Housing and Urban Development,

- RSF 5—Infrastructure Systems—U.S. Army Corps of Engineers, and
- RSF 6—Natural and Cultural—U.S. Department of the Interior.

The NDRF is divided into six recovery support functions (as stated above) that provide structure to facilitate problem solving, improve access to resources, and foster coordination among state, federal, nongovernmental partners, and other stakeholders. Each RSF has coordinating and primary federal agencies and supporting organizations that operate together with local, state and tribal government officials, nongovernmental organizations, and private-sector partners. The NDRF establishes three positions (listed below) that provide focal points for incorporating recovery considerations in the decision-making process and monitoring the need for adjustments in assistance wherever necessary and feasible. Those positions are federal disaster recovery coordinator, state or tribal disaster recovery coordinators, and local disaster recovery managers (FEMA, 2011b).

5.0 REFERENCES

Beaufort County. (2011). Beaufort County 2012–2012 disaster recovery plan. Beaufort County Council/Beaufort County Recovery Task Force: Beaufort County, South Carolina.

Beaufort County. (2012). Beaufort County disaster recovery operations policy plan. Beaufort County Council/Beaufort County Recovery Task Force: Beaufort County, South Carolina.

Crandall, W., & Spillman, J. E. (2010). *Crisis management in the new strategy landscape* (pp. 150–162). Sage Publications.

Cutter, S. L. (Ed.). (2001). *American hazardscapes: The regionalization of hazards and disasters.* Joseph Henry Press.

Cutter, S. L., Mitchell, J. T., & Scott, M. S. (2000). Revealing the vulnerability of people and places: A case study of Georgetown County, South Carolina. *Annals of the Association of American Geographers, 90*(4), 713–737.

Drabek, T. E., & Evans, J. (2010, June). Expanding our vision of emergency management through discussions of new teaching resources: The human side of disaster. Emergency Management Higher Education Conference, EMI, FEMA, DHS, Emmitsburg, Maryland.

Emergency Management Institute. (2010a). Integrated emergency management course IEMC-E930, Beaufort County, South Carolina, recovery planning. Federal Emergency Management Agency, Emergency

Management Institute. http://training.fema.gov/EMICourses/crsdetail.asp?cid=E930&ctype=R

Emergency Management Institute. (2010b). Recovery planning (presentation and exercise plan), integrated emergency management course IEMC-E930, Beaufort County, South Carolina. Federal Emergency Management Agency, Emergency Management Institute.

Federal Emergency Management Agency. (2005, December). *Long-term community recovery planning process: A self-help guide*. https://training.fema.gov/programs/emischool/el361toolkit/assets/long-termcommunity-recoveryplanningprocess.pdf

Federal Emergency Management Agency. (2008a). *Critical infrastructure and the national infrastructure protection plan*. http://emilms.fema.gov/IS821/SA0101020text.htm

Federal Emergency Management Agency. (2008b, January 1). *The national response framework, emergency support function #14—Long-term recovery annex*. http://www.fema.gov/pdf/emergency/nrf/nrf-esf-14.pdf

Federal Emergency Management Agency. (2009). *HAZUS, FEMA's methodology for estimating potential losses from disasters*. https://www.mmrs.fema.gov/plan/prevent/hazus/hz_esf14.shtm#3

Federal Emergency Management Agency. (2011a). *Critical infrastructure and key resources support annex*. https://www.cisa.gov/critical-infrastructure-and-key-resources-support-annex

Federal Emergency Management Agency. (2011b). *National disaster recovery framework*. http://www.emd.wa.gov/training/documents/NDRF_FEMA.pdf

Federal Emergency Management Agency. (2012a). *National incident management systems*. www.fema.gov/national-incident-management-system

Federal Emergency Management Agency. (2012b). *Public works and disaster recovery*. http://training.fema.gov/EMIWeb/IS/IS558.asp

Haddow, G., Bullock, J., & Coppola, D. P. (2008). *Introduction to emergency management* (3rd ed.). Butterworth-Heinemann.

Heinz, H. J., III. (2000). *The hidden costs of coastal hazards: Implications for risk assessment and mitigation*. Center for Science, Economics, and the Environment. http://www.csc.noaa.gov/cz/CZ07_Proceedings/PDFs/Wednesday_Abstracts/2825.Castellan.pdf

Leonard, H. B., & Howill, A. (2009). Chapter 2. In H. Kunreuther & M. Useen (Eds.), *Learning from catastrophes: Strategies for reaction and response*. Wharton School Publishing [Draft pre-publication, 2010]. www.hks.harvard.edu/var/ezp_site/storage/fckeditor/file/

May, J. (2008). Risk management. In S. Davies & H. A. Hertig (Eds.), *Security supervision and management* (3rd ed., pp. 243–246). Butterworth-Heinemann: Burlington.

Mohammed, J. (2008). Lull after the storm: Disaster management for utilities: Geographic information system plays a pivotal role in effective disaster management for the utilities. *SETLabs Briefings*, 6(2), 27–36. www.infosys.com/.../documents/disaster-management.pdf

Moore, T. L., Gulliver, J. S., Stack, L., & Simpson, M. H. (2016, October 1). Stormwater management and climate change: Vulnerability and capacity for adaptation in urban and suburban contexts. *Climatic Change*, *138*(3–4), 491–504.

Payne, J., Pyne, J., Berkowitz, S., & Clark, D. (2007). Surviving disastrous weather in small utilities. *Water and Environment Technologies, 19*(9), 1–2.

Phillips, B. D., & Neal, D. M. (2007). Recovery. In W. L. Waugh, Jr. & K. Tierney (Eds.). *Emergency management: Principles and practice for local government* (2nd ed., pp. 213–215). ICMA Press.

Schwab, J., Topping, K. C., Eadie, C., Deyle, R., & Smith, R. (1998). Earthquake case study: Loma Prieta in Santa Cruz and Watsonville, California. In *Planning for post-disaster recovery and reconstruction* (pp. 281–310). PAS Report 483/484. American Planning Association.

Smith, G. P. (2004). *Holistic disaster recovery: Creating a more sustainable future, online training*. Federal Emergency Management Agency, Emergency Management Institute. https://training.fema.gov/emiweb/downloads/hdr/emidrtsessionxiv%20final%209.17.04.pdf

Smith, G. P. (2010). *Planning for post-disaster recovery: A review of the United States disaster assistance framework*. P.E.R.I. Press.

Smith, G. P., & Wenger, D. (2006). Sustainable disaster recovery: Operationalizing an existing agenda. In H. Rodriguez et al. (Eds.), *Handbook of disaster research* (pp. 234–257). Springer.

U.S. Environment Protection Agency. (2007). *Decontamination and recovery planning: Water and wastewater utility case study* (pp. 1–13). (EPA 817-F/08-004; September 2008). U.S. Environmental Protection Agency, Office of Water. https://nepis.epa.gov/Exe/tiff2png.exe/P1001KL7.PNG?-r+75+-g+7+D%3A%5CZYFILES%5CINDEX%20DATA%5C06THRU10%5CTIFF%5C00000228%5CP1001KL7.TIF

U.S. Environment Protection Agency. (2008). *Water/wastewater WARN: Tabletop exercise facilitator guide* (p. 7). (EPA 817-R/08-005). nepis.epa.gov/Exe/ZyPURL.cgi?Dockey=P1001AF4.txt

U.S. Senate Committee on Governmental Affairs. (1993, March 31). Nomination of James Lee Witt to be Director, Emergency Management Agency, 101st Congress.

Webber, J., & Slaughter, N. (2010). Disaster recovery policies and procedures. In R. L. Kemp (Ed.), *Best practices for local government* (pp. 191–195). International City/County Management Association (ICMA).

6.0 SUGGESTED READINGS

Federal Disaster Recovery Programs: Brief Summaries. (2009, January 30). CRS report for Congress. Congressional Research Service, Library of Congress (RL31734). http://assets.opencrs.com/rpts/RL31734_20090130.pdf

Federal Emergency Management Agency. (n.d.). *Hazus*. https://www.fema.gov/flood-maps/products-tools/hazus

Federal Emergency Management Agency. (2005, June). *Long-term recovery assessment tool*. http://coscda.org/databases/COSCDA%20Disaster%20Recovery%20Toolkit-References-010209/Disaster%20Resources/DHS%20&%20FEMA/documents/fema-ltcr%20needs%20assessment%20tool.pdf

Federal Emergency Management Agency. (2010, March). *Public assistance applicant handbook*. https://www.fema.gov/pdf/government/grant/pa/fema323_app_handbk.pdf

Financing Recovery from Large-Scale Natural Disasters. (2009, February 9). CRS Report to Congress (RL34749). Congressional Research Service, Library of Congress. http://assets.opencrs.com/rpts/RL34749_20090209.pdf

Kousky, C., Luttmer, E. F. P., & Zeckhauser, R. J. (2006). Private investment and government protection. *Journal of Risk and Uncertainty, 22*(1), 73–100. http://ksghome.harvard.edu/~RZeckhauser/private_investment_and_government_protection.pdf

National Disaster Recovery Framework. (2018, September). http://www.fema.gov/pdf/recoveryframework/ndrf.pdf

National Infrastructure Protection Plan, Partnering to Enhance Protection and Resiliency. (2009). U.S. Department of Homeland Security. http://www.dhs.gov/xlibrary/assets/nipp_executive_summary_2009.pdf

Robert T. Stafford Disaster Relief and Emergency Assistance Act, as amended, and related authorities, FEMA 592. (2017). http://www.fema.gov/pdf/about/stafford_act.pdf

5

Resilience

1.0 RESILIENCE IN DISASTER MANAGEMENT 191
1.1 The Case for Resilience 191
1.2 Understanding Resilience in Disaster Management 192
1.3 Integrating Resilience in Prevention and Preparedness 196
1.4 Integrating Resilience Into Response and Recovery 196
2.0 RESILIENT INFRASTRUCTURE 199
2.1 Funding Resilient Infrastructure 202
2.2 Water and Flood Mitigation Infrastructure 202
2.3 Land-Use Planning 203
2.4 Design and Construction 203
3.0 A SYSTEMATIC APPROACH TO EMBEDDING RESILIENCE IN DISASTER MANAGEMENT 203
3.1 Integrating Resilience in Prevention and Preparedness 204
3.2 Integrating Resilience in Response and Recovery 205
3.2.1 Damage Assessments: DAMRsys™ 205
3.2.2 Betterment 207
3.3 Building Resilient Infrastructure 208
4.0 REFERENCES 210
6.0 SUGGESTED READINGS 213

1.0 RESILIENCE IN DISASTER MANAGEMENT

1.1 The Case for Resilience

Resilience is not a new concept. Humans have always been affected by disasters; sought means to preserve their lives, possessions and property; and then needed to recover and carry on. Resilience is something that can be defined as hidden capacities (physical, mental, economic, emotional, environmental) that we might not know we possess.

In current times, the economic, financial, social, and environmental effects of disasters can be great. A natural disaster event can mean that local economies suffer as business is disrupted and labor markets, supply chains, and consumption are affected. The financial costs, borne by communities, start with the cost of disaster response and continues into the long term as buildings, infrastructure, and other assets are repaired and replaced. The Australian Business Roundtable has estimated the total tangible and intangible economic costs of natural disasters in Australia at AU$18.2 billion per year, with this figure forecasted to rise to AU$39 billion per year by 2050 (Australian Business Roundtable, 2017, p. 20) Globally, similar data estimates are in the trillions.

Cost-benefit analysis shows investment in increasing resilience is more cost-effective in the long term than continuing "business as usual" (Australian Business Roundtable, 2016). While the effect of disasters on infrastructure, people, and the environment can be financially quantified, the lasting effects on mental and physical well-being and ecosystems are harder to quantify, but no less significant.

Disaster resilience is a long-term objective. Through collaboration, reprioritizing resilience measures, and facilitating community-led resilience, progress is achievable. Resilience in disaster management makes more sense when humans continually learn about the environment in which they live.

1.2 Understanding Resilience in Disaster Management

A comprehensive approach to disaster management comprises four phases: *prevention, preparedness, response,* and *recovery* (PPRR). This ensures a balance between the reduction of risk and the enhancement of community resilience, while ensuring effective response and recovery capabilities (Australian Business Roundtable, 2016). The PPRR activities focus on (Business Queensland, 2020):

Prevention: taking actions to reduce or eliminate the likelihood or effects of an incident;

Preparedness: taking steps before an incident to ensure effective response and recovery;

Response: containing, controlling, or minimizing the effects of an incident; and

Recovery: taking steps to minimize disruption and recovery times.

Resilience must be embedded in all of these phases of the traditional disaster management cycle (Figure 5.1). In particular, resilience measures must be incorporated into prevention and preparation before a disaster occurs, rather than solely in response and recovery operations. Building this disaster resilience is not only the domain of emergency management agencies; it

FIGURE 5.1 Incorporating Resilience Into the PPRR Model

is a shared responsibility across society as a whole (Council of Australian Governments, 2011).

In the context of disaster management, resilience can be defined as *a system or community's ability to rapidly accommodate and recover from the impacts of hazards, restore essential structures and desired functionality, and adapt to new circumstances* (Queensland Reconstruction Authority [QRA], 2017, p. 9). The concept of resilience also offers transformational opportunities and can be used to identify what can be maintained, what needs to be modified, and what must be transformed to reduce disaster risks to the community (O'Connell et al., 2020).

National and international experience has shown that resilient organizations or communities have risk-informed and appropriately prepared individuals; the capacity to adapt; and healthy levels of community connectedness, trust, and cooperation.

The resilience of individuals, communities, infrastructure, and the natural environment are interlinked (Figure 5.2). A community is only as resilient as the collective resilience of individuals in the community. Individuals can only be resilient when they have resources, information about risks, and the support of their community networks. The resilience of a community is also directly influenced by the resilience of the community's infrastructure, as this can affect emergency response, evacuation, and access to basic needs such as shelter and fresh water. Ecosystems and natural assets also mitigate against natural hazards. However, modified environments are driving climate change and more severe natural disasters, so damage to the environment has consequences for future resilience. Therefore, it is essential to protect the environment to maintain its functions and reduce the severity of disasters.

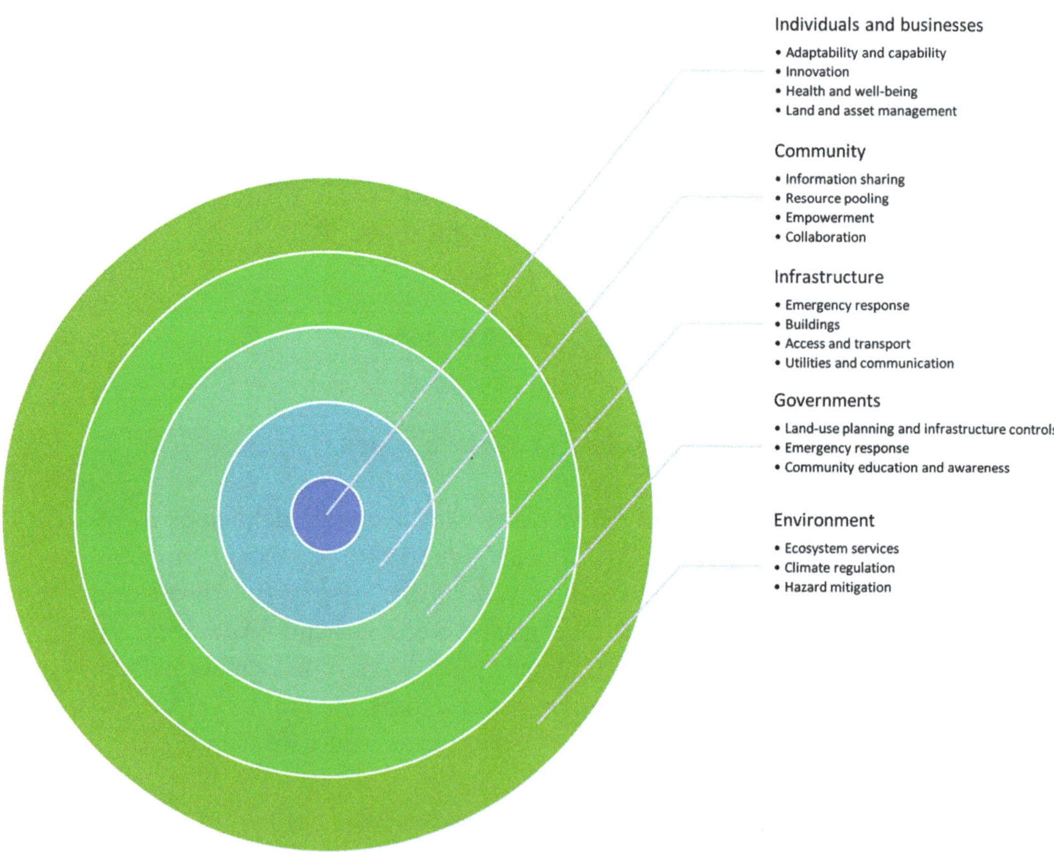

FIGURE 5.2 Interdependencies That Influence Resilience

All levels of our society and environment contribute to resilience, but all have different needs to achieve it. Therefore, collaboration is required to achieve the right balance.

The evidence that human changes to the landscape exacerbate disaster damage is overwhelming. It is also clear that simply building structural defenses is not always cost-effective and can simply shift the problem. Around the world, interest on ecosystem-based approaches to natural hazard mitigation is increasing (Queensland Department of Environment and Heritage Protection, 2012).

The natural environment also provides "ecosystem services" for which costs are incorporated into economic models to demonstrate the broader benefits of protecting and managing the environment, and support investment and decision-making. Ecosystem services include (Comberti et al., 2015):

Provisioning services: Products obtained from ecosystems, such as food, fresh water, fuel, and materials;

Regulating services: Benefits obtained from regulation of ecosystem processes, such as climate regulation, water purification, air quality regulation, flood protection, and erosion control; and

Cultural services: Nonmaterial benefits obtained from ecosystems, such as recreation, tourism, education, identification, cultural heritage, health, and well-being.

To maintain these benefits, there are reciprocating services that must be provided to ecosystems in return (Comberti et al., 2015):

Protecting services: Actions that protect ecosystems, such as habitat protection and weed management;

Enhancing services: Actions that modify or extend ecosystem processes, such as cultivation, fertilization, and nutrient (re)cycling; and

Restoring services: Actions that complement ecosystem processes, such as improving soil, air, and water quality; habitat restoration; nutrient release (burning); planting and harvesting.

We need to better understand how our landscapes function and what we need to do to absorb the energy of weather extremes. Landscape processes are complex, but there are measures that can be incorporated into landscaping, natural resource management, and environmental planning to mitigate disasters. It is essential to plan a whole catchment or regional approach based on a good understanding of that particular catchment or region's processes (Queensland Department of Environment and Heritage Protection, 2012). There is a strong case for better understanding and harnessing the benefits of natural assets for resilience.

1.3 Integrating Resilience in Prevention and Preparedness

Research tells us that the most powerful action to address disaster risk can be taken at the community level. Communities play an active and central role in disaster risk prevention and preparedness. They are the first to experience rapid changes and the first to be called upon to adapt following a disaster.

Resilience building must be locally led and must use local disaster risk knowledge. Resilience is significantly increased by active planning and preparation for protecting life, infrastructure systems, the natural environment, and property, based on an awareness of the disaster risks relevant to a specific locality (Council of Australian Governments, 2011).

The concept of resilience to natural hazards is intrinsically linked to the things communities value and how they may cope with the possible loss of those things of value. As people and communities are inherently motivated to protect what they value, recovery and resilience planning that places communities' needs and values at the core will ultimately deliver better outcomes.

As well as being local, resilience building needs to be long term and continual. Critically, activities to improve levels of resilience to disasters rely on changing how we think and operate as part of our "business as usual." Building resilience must be an ongoing process of learning, adjustment, and adaptation that starts well before, and continues well beyond, response and recovery.

Building resilience into emergency planning also requires a thorough understanding of risk and risk management processes. Risk management is a proactive process that forms the basis for disaster management planning and helps those responsible for disaster management to provide effective, relevant, and informed services to communities (Queensland Government, 2020, Section 3.2).

Table 5.1 provides an indication of the actions required by different stakeholders to improve resilience. It also highlights the interdependencies of these actions, based on the capacity and capability of each stakeholder.

1.4 Integrating Resilience Into Response and Recovery

The need for recovery may arise from a range of disaster events, encompassing natural and non-natural disasters such as floods, cyclones, bushfires, acts of terrorism, and major health emergencies.

Following a disaster, individuals, families, groups, organizations, and communities will have differing needs—some may be directly affected by the event through injury, death, or loss of property, while others may be evacuated or isolated, emotionally affected, or financially affected through

TABLE 5.1 Roles and Responsibilities in Building Resilience Into Prevention and Preparedness

People and businesses	Communities	Emergency sector	Governments
• Stay informed about hazards and risks • Engage with local communities, emergency services, and governments about how to prevent and prepare for disasters • Take steps to prevent or mitigate risks of hazards to themselves, their assets, and their livelihoods, including homes, possessions, land, local environment, cultural heritage, and economic capital • Have up-to-date plans in place for responding to a disaster • Be prepared for known risks, and be ready to adapt to emerging risks and become self-reliant • Be connected to receive alerts and warnings of forecast events	• Facilitate social cohesion • Establish and maintain social support systems, community partnerships, and networks • Understand who can provide help, and who may need help, including those who may be exposed or vulnerable • Share information and resources • Have community plans in place for responding to a disaster, with clear actions, roles, responsibilities, and built-in flexibility to adapt plans to changing circumstances • Contribute to ongoing resilience learning	• Build resilience into prevention and preparedness planning and activities • Enable and contribute to ongoing resilience learning • Incorporate learning from previous disasters and studies into prevention and preparedness planning and activities	• Undertake and share risk assessments and mitigation options • Provide guidance on how communities and individuals can prevent and prepare for events • Implement land-use planning and building controls that prevent exposure to hazards and mitigate risks • Facilitate and provide resilient infrastructure • Establish and maintain communication networks • Increase education and awareness • Facilitate individual and community-led resilience • Enable and contribute to ongoing resilience learning

loss of employment or livelihood. Effects of a disaster on the natural environment may include damage or loss of flora and fauna, poor air quality, reduced water quality, land degradation and contamination, as well as heritage-listed place issues (QRA, 2017). Given the wide-ranging effects, the response to a disaster can require multiple agencies and the community to work together at local, district, and state levels to deliver recovery outcomes (QRA, 2017, p. 10).

Community-led recovery approaches are effective in supporting the rapid restoration of services essential to human well-being and present an opportunity to build resilience and improve a community's circumstances and preparedness beyond its pre-disaster status. Community-led recovery requires a degree of flexibility and understanding from the governments that have established the laws and rules for society. Governments should acknowledge that the "business as usual" rules may not apply in disaster recovery, as communities work toward securing their safety, accessing basic services, and regaining a sense of normality. That said, governments also have a responsibility to ensure that actions taken by communities do not result in perverse outcomes that create further hazards. Well-meaning people may unintentionally make situations worse, if they are not aware of all risks and do not have sufficient, accurate, and up-to-date information. An example of this is inexperienced people reconnecting damaged water supplies or repairing roads to facilitate access. The potential for health effects from contaminated water or reduced vehicle safety can be high. But governments also need to be mindful that communities know their local area best and will be able to pool resources and develop innovative solutions, particularly where government resources are stretched or otherwise not available.

Community-led adaptation will depend on the community, location, natural environment, and consequences of the disaster event. The key to success in community-led adaptation is facilitating flexibility, making options available, sharing ideas, identifying barriers (such as cultural norms or past experiences), and setting boundaries on what adaptation approaches might be acceptable or unacceptable. These will drive acceptable behavior change and positive outcomes that are safe for the community and aid recovery. Governments should be ready to empower communities to take action to adapt to changing circumstances for quicker recovery within acceptable boundaries to manage risks.

In the case of environmental recovery, restoration may require a long-term strategy to re-establish self-sustaining ecosystems. Environmental recovery requires local input, expert knowledge, and committed resources, and may include regeneration of the natural environment, associated infrastructure and heritage sites and structures, and the management of pollution and contamination (QRA, 2017). Environmental recovery from disasters can be expensive; therefore, prioritizing reciprocating services (protection, enhancement, and restoration) at the planning and preparedness stages can mitigate the effects of disasters and enable recovery resources to be redirected to other recovery actions.

Resilience through recovery is predicated on earlier resilience-building measures through the prevention and preparedness stages of disaster management. Notably, the existence of elements of resilience in communities before a disaster is essential to building resilience in recovery (QRA, 2017).

While short-term recovery responses are vital, there is also a need to foster longer-term resilience, and reduce industry, community, and environmental exposure to natural disasters (Commonwealth Scientific and Industrial Research Organisation [CSIRO], 2020). Internationally, qualitative analyses of country experiences in disaster recovery undertaken by the Global Facility for Disaster Reduction and Recovery (2015) indicate that there are four progressive stages for a country to successfully integrate disaster risk reduction into disaster recovery measures:

1. Integrating disaster risk reduction into recovery needs assessments,
2. Sustained commitment to disaster risk reduction during recovery planning,
3. Incorporating disaster risk reduction in the design and implementation of recovery programs, and
4. Translating the gains of resilient recovery into sustainable development.

Planning for recovery and reconstruction following a natural disaster can often provide a unique opportunity to rebuild stronger, more resilient communities, economies, and natural and built environments. Post-disaster, a "window of opportunity" emerges where resilience improvements can be made more readily, given the focus on—and commitment to—the recovery effort.

2.0 RESILIENT INFRASTRUCTURE

Infrastructure underpins every aspect of modern life. It supports communities and facilitates productivity and resilience. Infrastructure systems provide services that generate economic and social prosperity such as energy, food, water, transport, communications, health, and banking and finance.

Functionality and resilience of infrastructure directly influence disaster responses, such as evacuation and rescue as well as recovery and reconstruction (Gajanayake et al., 2018). The overlapping nature of contemporary infrastructure systems means that several services depend on each other to function effectively. This interconnectedness and interdependency places entire infrastructure systems at risk of disruption when a natural hazard hits.

Loss of, or damage to, critical infrastructure can affect supply security and service continuity, causing severe disruption to business, governments, and the community. For example, flood damage to highways can disrupt critical agricultural supply chains, and damage to port facilities caused by cyclones or hurricanes can affect international trade. Therefore, the resilience of infrastructure that individuals, businesses, and communities rely on—not only for business as usual, but also for rapid disaster response and recovery—must be front and center in disaster prevention, preparedness, response, and recovery. For example:

- Roads, bridges, and other transport infrastructure enable access and evacuation during disaster events, but are also necessary for businesses to function, and for people to access work and education. Resilient transport infrastructure that can better withstand disasters will enable communities to recovery faster. Transport corridors are also often used for other utilities, so their resilience is essential for maintaining people's access to information and resources.
- Electricity and fuel are a key component of society, enabling transport, use of communications equipment, safe food storage, lighting, and life-support devices. These become even more crucial during disasters, when people and communities are potentially isolated for extended periods. Secure supplies help to sustain people during response and recovery.
- Communications infrastructure allow connections and sharing of information by phone, internet, mobile devices, and radio. Data flows are a core element of our societies and economies, and information availability enables faster and targeted disaster response.
- Water supply is a basic need. Access to fresh water is essential for life, both in disasters and during business as usual. Supplies that are cut or contaminated during disasters place people at risk.

Resilient infrastructure is planned, designed, and built to withstand disruption. It operates effectively in crisis and deals with and adapts to shocks and stresses. Resilient infrastructure can adapt to present and future conditions, remain functional during a disaster, or restore functionality quickly following a disaster event. Resilient infrastructure allows essential services to continue or quickly return to service, ensuring that a community's economic productivity, health, and well-being are safeguarded during and after a natural disaster.

Despite the high capital and ongoing maintenance costs often associated with resilient infrastructure, investment yields significant savings in avoided costs. For example, economic benefits can be measured in terms of reductions in future flood damage, for both tangible and intangible damage. Therefore, the feasibility of new infrastructure for mitigation purposes should not be judged solely on financial merit. It should also consider a range of social and environmental factors, including potential benefits to integrated catchment planning outcomes. Assessments can be used to evaluate and compare the resilience merits of various structural options across multiple criteria including community safety, economic viability, technical feasibility, community attitudes, infrastructure and transport, environmental effects, and natural resource management. Employing multiple-criteria assessments in the infrastructure planning stage is a critical tool to enable resilience building (QRA, 2019).

An investment framework, which is collaboratively developed and adopted, can be a useful tool to provide consistent guidance on effective investment decision-making and prioritization.

Strengthening a public asset so that it remains operational during a natural disaster, or is non-operational for a shorter period, is central to enhancing a community's resilience. It also increases public safety and minimizes economic effects during and after disaster events. The Australian Business Roundtable (2016) identifies exposure, vulnerability, and effect as three key areas to strengthen resilience:

- Reducing the infrastructure's *exposure* to disaster hazards: For example, relocating infrastructure away from areas susceptible to hazards, such as roads in areas less prone to flooding;
- Reducing the infrastructure's *vulnerability* to disaster hazards: For example, changing infrastructure design or materials to reduce the severity of effects, such as shifting transmission lines underground in areas prone to bushfires;
- Reduce the *effect* of disaster hazards on infrastructure: For example, introducing early warning, evacuation, and/or contingency systems for emergency responses during service losses, such as informing customers of expected network outages.

Strategies to mitigate risk, and reduce exposure and vulnerability to and effect of natural disasters, include (Queensland Government, 2020, section 3.4)

- community awareness campaigns to increase knowledge of how to prepare for disaster events,
- community education programs to build knowledge of the appropriate actions to prepare for and respond to a disaster event,
- hazard specific control activities such as flood levees or bushfire mitigation strategies,

- design improvements to infrastructure or services
- capital works such as levee bank construction to reduce the effects of flooding,
- resilience activities including partnerships and engagement between sectors, and
- annual programs (e.g., vegetation management around essential services and essential infrastructure such as power lines).

A coordinated approach to building resilience in infrastructure requires resilience to natural disasters being considered in policy and investment frameworks, land-use planning, and design and construction. This integrated approach will allow consideration of cross-disciplinary opportunities and risks, and will help achieve multiple benefits for resilience across the interdependencies of infrastructure, individuals, communities, and the natural environment.

2.1 Funding Resilient Infrastructure

It is internationally recognized that upfront investment in more resilient infrastructure saves money for all levels of government over the long term and reduces the community effect of disasters by saving lives and producing additional co-benefits (Deloitte Access Economics, 2013). A study of 5500 mitigation grants approved by the U.S. Federal Emergency Management Agency (FEMA) reported an overall benefit-cost ratio of 4:1 between 1993 and 2003; approximately $4 saved for every $1 spent on mitigation (Rose et al., 2007).

Increased funding for "building back better" or "betterment" projects enables essential public assets to build to a more resilient standard post-disaster, improving the ability of infrastructure to withstand the effects of future disasters. The concept of betterment, often considered in post-disaster recovery and reconstruction, should be a fundamental pre-disaster consideration. It should be pursued through proactive mitigation strategies that aim to enhance and harden infrastructure to a more disaster-resilient standard (Queensland Government, 2020, section 3.4).

2.2 Water and Flood Mitigation Infrastructure

Infrastructure located within floodplains can modify flood behavior. Flood and water infrastructure generally involve heavily engineered works and can encompass a range of structural options for flood risk management, including dams, levees, detention basins, and flood gates. As riverine flood behavior is sensitive to flow and floodplain conditions, infrastructure can be used to alter this behavior and potentially mitigate adverse impacts (QRA, 2019).

Central to a dam's ongoing flood mitigation effectiveness is the ability to establish flood operation procedures that reduce adverse effects downstream. Planning for the reduction of flood effects requires an understanding of potential damages across the range of flood flows (QRA, 2019).

2.3 Land-Use Planning

Using a risk-based approach to land-use planning, taking into account natural hazard risks and assigning land uses appropriate to that level of risk are effective means of improving infrastructure resilience to all hazard events and future risks.

To make well-informed decisions, infrastructure planners must have ready access to data and information to determine potential risks posed by hazards at a site-specific level. Infrastructure planning and design should be considered through a resilience lens and with a partnered approach to working with the community, end users, and other agencies.

2.4 Design and Construction

Making infrastructure more resilient through the design and construction phase contributes to savings across the entire infrastructure life cycle, especially when infrastructure is affected after a disaster, when less time and money is spent on re-establishing services and reconstruction. This is true across all levels of infrastructure, from critical community infrastructure, to building resilient households.

Resilient building design can significantly reduce the effort and time needed to return people to their homes and workplaces following natural disasters. For homeowners, resilient design and construction can also reduce long-term costs associated with disaster damage and insurance premiums, as well as the social and emotional effects of disasters.

3.0 A SYSTEMATIC APPROACH TO EMBEDDING RESILIENCE IN DISASTER MANAGEMENT

Queensland is the most disaster-affected state in Australia and has been hit by more than 80 natural disaster events in the past decade. The Queensland Reconstruction Authority was established following wide-ranging flooding over the Australian summer of 2010–2011 and is Australia's first permanent disaster recovery agency.

The QRA has been instrumental in developing policy frameworks and tools that consider resilience across the disaster management cycle:

- The Queensland Strategy for Disaster Resilience and its implementation plan, Resilient Queensland, commit to delivering more resilient infrastructure and incorporating risk reduction into planning and development.
- The Queensland Disaster Resilience and Mitigation Investment Framework supports prioritizing, and investing in, resilience and mitigation across Queensland.
- Queensland's Resilience Checklist is an analytical tool for considering whole-of-system interventions that can be applied to build greater resilience into infrastructure planning processes.
- The Queensland Flood Risk Management Framework emphasizes a risk-based, multidisciplinary, catchment-wide approach to building community resilience to flooding.

3.1 Integrating Resilience in Prevention and Preparedness

To ensure effective disaster risk management, emergency planners at all levels are encouraged to

- Undertake risk assessment and management using an approved, recognized methodology that considers all reasonably foreseeable hazards, both natural and human-made;
- Support risk assessment and management with scientific data, the use of geospatial information systems, and analysis of historical and/or projected effects to identify area-specific exposures and vulnerabilities;
- Identify residual risk (the risk that remains in unmanaged form, even if controls are in place) and develop strategies to either accept, mitigate, or transfer that risk; and
- Clearly document and make publicly available hazard identification and risk assessments to stakeholders and community members, and review them regularly.

The Queensland Emergency Risk Management Framework is used to inform disaster management planning at all levels of Queensland's disaster management arrangements. The Framework applies a standardized and internationally recognized approach to the prioritization, mitigation, and management of risk to directly inform planning and resource allocation, and to promote active communication, cooperation, and coordination.

Use of the framework can inform the regional resilience strategies that are being developed across the state as part of Resilient Queensland (Figure 5.3). These strategies are being developed to identify improvements to vital infrastructure critical to public safety and the economies of the regions. To date, three regional resilience strategies have been developed, and by 2022, all 77 local governments in Queensland will be part of a regional resilience strategy that clearly identifies and prioritizes actions to strengthen disaster resilience.

At the household and community level, the annual Get Ready Queensland campaign (Queensland Government, n.d.) provides individuals and Queensland local governments with information and resources to promote all-hazards preparedness activities.

3.2 Integrating Resilience in Response and Recovery

3.2.1 Damage Assessments: DAMRsys™

The collection of spatial and impact data about damage from natural disaster events informs targeted recovery assistance. It also helps to identify vulnerabilities in infrastructure networks. Conducting timely post-event impact assessments that document infrastructure damage gives decision makers the

FIGURE 5.3 Resilient Queensland (Source: QRA, 2018)

information they need to develop appropriate disaster recovery programs and to invest in more resilient infrastructure.

Queensland's ability to recover from natural disasters has been significantly enhanced with the development of system to gather early impact information through initial damage assessments, as well as follow-up assessments throughout the recovery process (Figure 5.4). The Damage Assessment and Reconstruction Monitoring system (DARMsys™) is used to monitor Queensland's rebuilding progress.

Real-time data is collected by officers from QRA and Queensland Fire and Emergency Services (QFES), who undertake damage assessments in the community. Damage assessors use a hand-held monitoring device to collect map-based damage data from street-by-street and house-by-house inspections through communities affected by flooding, bushfires, and cyclones. The data provide valuable intelligence on the level of damage, damage patterns, and where the greatest recovery needs exist.

QRA also conducts periodic audits and quarterly follow-ups on this impact data to monitor recovery processes and determine ongoing needs. DARMsys™ information has been crucial to identifying the scale and scope of disaster effects and is now used by multiple Queensland government

FIGURE 5.4 Conducting In-Field Damage Assessments (Source: QRA, n.d.-a)

agencies to plan their response and recovery activities. DARMsys™ is acknowledged by the World Bank (2011) as having played an instrumental role in enabling Queensland to recover quickly from natural disasters.

QRA has also developed the Repeat Event Dollars Index (REDI), which is a geospatial database that captures all road infrastructure assets damaged following disaster events (Figure 5.5). As of November 2020, QRA has mapped almost 10 years of damage data including AU$5.5 billion in reconstruction costs. REDI information is made available to local councils, at no cost, to provide a better understanding of their disaster risk and to help identify priority works. Councils can also use REDI to make informed decisions regarding future investment in stronger and more resilient infrastructure through infrastructure betterment programs (QRA, n.d.-b).

3.2.2 Betterment

Since 2013, Queensland's Betterment Fund has demonstrated a commitment by all levels of government in Australia to work together to improve community resilience.

Following Severe Tropical Cyclones Oswald, Marcia, and Debbie in 2013, 2015, and 2017 respectively, the Queensland Betterment Fund has committed AU$141.85 million for 370 projects across 44 local government areas. The Betterment Fund allows locals governments to rebuild essential

FIGURE 5.5 Geospatial Overlay of Damage Assessment Data in Townsville, Queensland, Following North and Far North Queensland Monsoon Trough Event, 2019 (Source: QRA)

public assets to a more resilient standard that helps them withstand the effects of future disasters. To date, 260 of the total 370 betterment projects delivered have been subjected to subsequent natural disasters since their completion, with some sites subjected up to five times by a combination of floods, cyclones, and severe storms.

From an investment of approximately AU$93 million for those 260 assets subjected to subsequent events, more than AU$145 million in avoided costs have been realized. A further AU$100 million Betterment Fund is currently being delivered following the North and Far North Queensland Monsoon Flood in January and February 2019.

The Gayndah Water Supply Intake Station is a betterment case study (Carroll, 2015). Gayndah is a rural town approximately 320 km (200 miles) northwest of Brisbane, Queensland. An agricultural area, it is the nation's second biggest exporter of citrus and produces much of Australia's mandarin orange crop.

The Gayndah Water Supply Intake Station on the Burnett River provides the town's only water supply and supports a population of approximately 2000, as well as local primary industries including dairy farming. The North Burnett Regional Council estimates AU$3.85 million in damage occurred to the asset following 3 years of disaster events. The pumping station was severely damaged in 2011 and was rebuilt at a cost of AU$1.22 million.

These works were completed just before Severe Tropical Cyclone Oswald in early 2013, which subsequently destroyed the new jetty, raw water intakes, pumps, and water discharge main. Reconstruction of the infrastructure in its existing position would not provide resilience to future natural disaster events and would subject the community's water supply to ongoing risk.

The council's Betterment project for the Gayndah Water Supply included a new submersible-style pumping station, a water intake upstream of the weir, and a new raw-water main to the water treatment facility. The cost of relocating the Gayndah Water Supply Intake was AU$3.95 million (Figures 5.6 and 5.7). Since completion of these works, the Gayndah Water Supply Intake has been functional through two significant events, Tropical Cyclone Marcia in 2015 and the 2016 Central Queensland Surface Trough. Avoided reconstruction costs from these two events have been on the order of AU$5.4 million.

3.3 Building Resilient Infrastructure

Cyclones, floods, and storm tides are a part of life for many Queenslanders, and building resilience at the household level is essential to ensure that communities can recover quickly from natural disasters.

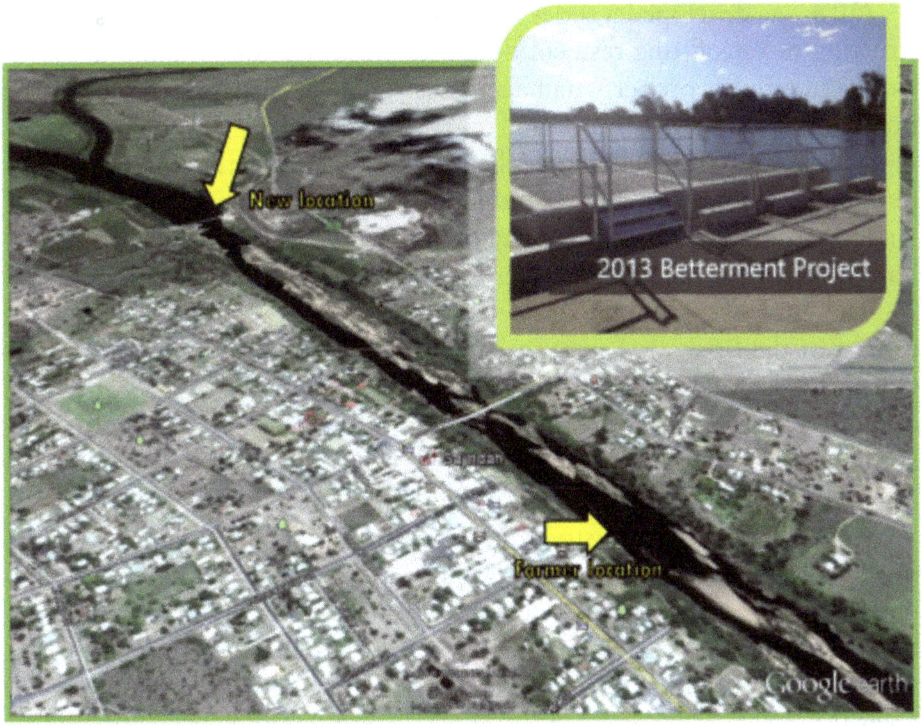

FIGURE 5.6 Image Showing the Relocation of the Gaydah Water Intake to Above the Claude Wharton Weir. (Inset picture shows the completed 2013 Betterment Project, which included a new submersible style pumping station and a new raw-water rising main to the water treatment facility) (Source: QRA)

FIGURE 5.7 Flooding Effects of the 2015 Flood on the Gayndah Water Supply Intake (Source: QRA)

QRA has developed a series of guidelines to assist homeowners in constructing or retrofitting resilient designs, which improves community resilience and reduces property damage. These guidelines cover cyclones, storm tides, and floods.

A case study for building resilient infrastructure is the Flood Resilient Building Guidance for Queensland Homes, a joint initiative of the Queensland Government, Brisbane City Council, Ipswich City Council, Lockyer Valley Regional Council, Somerset Regional Council, and Seqwater.

The guidance is one of a suite of flood resilience initiatives developed as part of the Brisbane River Catchment Flood Studies. It responds to the Queensland Floods Commission of Inquiry's (2012) *Final Report*, which called for an appropriate mix of measures including building controls to minimize the effects of floods and help to reduce the cost of property damage and time taken to restore a building after a flood.

Aimed at building industry professionals, state and local authorities, and owners of residential properties in flood-prone areas across Queensland, this nonmandatory guidance may be considered in addition to the mandatory requirements contained in MP3.5 of the Queensland Development Code, which specifically addresses the construction of buildings in flood hazard areas (Queensland Department of Housing and Public Works, 2013).

The guidance provides detailed information on flood-resilient construction systems, building materials, and products, and the economic benefits and cost comparisons of regular and flood-resilient construction.

4.0 REFERENCES

Australian Business Roundtable. (2016). *Building resilient infrastructure*. http://australianbusinessroundtable.com.au/our-research/resilient-infrastructure-report#:~:text=In%20March%202016%2C%20the%20Australian,released%20'Building%20Resilient%20Infrastructure'.&text=The%20report%20is%20based%20on,major%20casualty%20of%2C%20natural%20disasters

Australian Business Roundtable for Disaster Resilience & Safer Communities. (2017). *Building resilience in our states and territories*. Australian Business Roundtable.

Business Queensland. (2020). The PPRR risk management model. https://www.business.qld.gov.au/running-business/protecting-business/risk-management/pprr-model

Carroll, F. (2015). Building it back better to reduce risks after multiple disasters. [Paper presentation]. Floodplain Management Association National

Conference, Brisbane, Australia. https://www.floodplainconference.com/papers2015/Frankie%20Carroll%20Full%20Paper.pdf

Comberti, C., Thornton, T. F., Wyllie de Echeverria, V., & Patterson, T. (2015). Ecosystem services or services to ecosystems? Valuing cultivation and reciprocal relationships between humans and ecosystems. *Global Environmental Change*, *34*, 247–262. https://doi.org/10.1016/j.gloenvcha.2015.07.007

Commonwealth Scientific and Industrial Research Organisation (CSIRO). 2020. *CSIRO Report on Climate and Disaster Resilience*. https://www.csiro.au/en/Research/Environment/Extreme-Events/Bushfire/frontline-support/report-climate-disaste-resilience

Council of Australian Governments. (2011). *National strategy for disaster resilience: Building the resilience of our nation to disasters*. https://www.homeaffairs.gov.au/emergency/files/national-strategy-disaster-resilience.pdf

Deloitte Access Economics. (2013, June). *Building our nation's resilience to natural disasters*. Australian Business Roundtable for Disaster Resilience and Safer Communities. http://australianbusinessroundtable.com.au/assets/documents/ABR_building-resilience-in-our-states-and-territories.pdf

Gajayanake, A., Mohseni, H., Zhang, G., Mullet, J., & Setunge, S. (2018). Community adaptation to cope with disaster-related road structure failure. *Procedia Engineering, 212,* 1355–1362. https://doi.org/10.1016/j.proeng.2018.01.175

Global Facility for Disaster Reduction and Recovery. (2015). *Resilient recovery: An imperative for sustainable development*. https://www.preventionweb.net/files/44171_resilientrecoveryanimperativeforsus.pdf

O'Connell, D., Grigg, N., Hayman, D., Bohensky, E., Measham, T., Wise, R., Maru, Y., Dunlop, M., Patterson, S., Vaidya, S., & Williams, R. (2020). *Disaster-resilient and adaptive to change: Narratives to support co-ordinated practice and collective action in Queensland*. Commonwealth Scientific and Industrial Research Organisation.

Queensland Department of Environment and Heritage Protection. (2012). *Natural assets for flood and cyclone resilience*. https://wetlandinfo.des.qld.gov.au/resources/static/pdf/management/synthesis-scientific-evidence.pdf

Queensland Department of Housing and Public Works. (2013). *Queensland Development Code MP 3.5: Construction of buildings in flood hazard areas*. www.hpw.qld.gov.au/construction/BuildingPlumbing/Building/

BuildingLawsCodes/QueenslandDevelopmentCode/Pages/QueenslandDevelopmentCodeCurrentParts.aspx

Queensland Floods Commission of Inquiry. (2012). *Final report*. www.floodcommission.qld.gov.au/publications/final-report/

Queensland Government. (n.d.). Get Ready Queensland. Retrieved November 1, 2020, from https://www.getready.qld.gov.au/

Queensland Government. (2020). Disaster management. https://www.disaster.qld.gov.au/dmg/Pages/DM-Guideline.aspx#3

Queensland Reconstruction Authority. (n.d.-a). Damage Assessment and Reconstruction Monitoring (DARMsys). Retrieved November 1, 2020, from https://www.qra.qld.gov.au/key-projects/damage-assessment-and-reconstruction-monitoring-darmsystm

Queensland Reconstruction Authority. (n.d.-b). Repeat events and dollars index application. Retrieved November 1, 2020, from https://www.qra.qld.gov.au/REDI

Queensland Reconstruction Authority. (2017). *Queensland recovery plan*. https://www.qra.qld.gov.au/sites/default/files/2019-08/Queensland-Recovery-Plan-update-August2019.PDF

Queensland Reconstruction Authority. (2018). *Resilient Queensland 2018–21: Delivering the Queensland strategy for disaster resilience*. https://www.qra.qld.gov.au/sites/default/files/2018-10/resilient-queensland-2018-21-final_0.pdf

Queensland Reconstruction Authority. (2019, February). *Brisbane River strategic floodplain management plan*. https://www.qra.qld.gov.au/sites/default/files/2019-04/Plan%20-%20Brisbane%20River%20Strategic%20Floodplain%20Management%20Plan%20%28April%202019%29_2.pdf

Rose, A., Porter, K., Dash, N., Bouabid, J., Huyck, C., Whitehead, J., Shaw, D., Eguchi, R., Taylor, C., McLane, T., Tobin, L., Ganderton, P., Godschalk, D., Kiremidjian, A., Tierney, K., & West, C. (2007). Benefit-cost analysis of FEMA hazard mitigation grants. *Natural Hazards Review, 8*(4), 97–111.

World Bank. (2011). *Queensland: Recovery and reconstruction in the aftermath of the 2010/11 flood events and Cyclone Yasi*. http://qldreconstruction.org.au/u/lib/cms2/world-bank-report-full.pdf

6.0 SUGGESTED READINGS

Adams, T. M., & Anderson, L. R. (2019). *Policing in natural disasters: Stress, resilience, and the challenges of emergency management*. Temple University Press.

Busch, N. E., & Givens, A. D. (2013). Achieving resilience in disaster management: The role of public-private partnerships. *Journal of Strategic Security*, 6(2), 1–19.

Herwig, A., & Simonici, M. (Eds.). (2019). *Law and the management of disasters: The challenge of resilience*. Routledge.

Saja, A. M. A., Teo, M., Goonetilleke, A., Ziyath, A. M., & Gunatilake, J. (2020). Selection of surrogates to assess social resilience in disaster management using multi-criteria decision analysis. *International Journal of Disaster Resilience in the Built Environment, 11*(4), 453–480.

Surminski, S., & Tanner, T. (Eds.). (2016). *Realizing the triple dividend of resilience: A new business case for disaster risk management*. SpringerLink.

Index

A
American National Standards Institute (ANSI), 144
American Society of Civil Engineers, 144
American Water Works Association (AWWA), 134
ANSI. *See* American National Standards Institute
ASDWA. *See* Association of State Drinking Water Administrators
Asset
  definition of, 100
  public, strengthening, 201
Association of State and Territorial Health Officials (ASTHO), 134
Association of State Drinking Water Administrators (ASDWA), 134
ASTHO. *See* Association of State and Territorial Health Officials
Avalanches, 13
AWWA. *See* American Water Works Association

B
Beyond-localized emergency (high-level emergency), 108
Biological disasters, 11–12
Blank media query log sheet, 104

C
Case studies, 7
CDC. *See* Centers for Disease Control and Prevention
Centers for Disease Control and Prevention (CDC), 98–99
Chemical, biological, radiological, nuclear, and high-yield explosive (CBRNE) threats, 98
Chlorine gas, 20
CIPAC. *See* Critical Infrastructure Partnership Advisory Council
Civil unrest, 19
Climate Resilience Evaluation and Awareness Tool (CREAT), 97, 142
Climatological disasters, 6, 18
  drought, 17–18
  dust storms, 17
  extreme temperatures, 17
  wildfires, 17
Code of Federal Regulations (CFR), 20
  29 CFR 1910.38, 40, 66
  29 CFR 1910.39, 40
  29 CFR 1910.119, 40
  29 CFR 1910.1200(9), 8
  40 CFR 68.150-195, 40, 44
  40 CFR 109, 42
  40 CFR 109.1-6, 40
  40 CFR 112.1-12, 41
  40 CFR 112.3-12, 40
  40 CFR 112.20, 39–40
  40 CFR 122.26, 40, 42
  40 CFR 264.50-56, 40, 43
  40 CFR 280, 41
  40 CFR 281, 41
  29 CFR 1910.39(c)(1), 70, 71
  29 CFR 1910.39(c)(2), 71
  29 CFR 1910.39(c)(4), 71
Colorado Discharge Permit System, 103
Communication, 9–11, 25–26, 125–133
  additional useful tips for, 132–133
  after determininng problem, 128
  avoid saying "no comment," 129
  continuity planning, 9–10
  express concern for victims, 128–129
  holding statement, 126, 128
  identify affected parties, 129–130
  information for response procedure, 125
  infrastructure, 200
  intra-agency, 183
  media briefing, conducting, 131
  during recovery, 182–183
  respond to questions of media, 129

taking photographs, 131–132
technological failures, 10
training, 102–103
two-way conversations, preparation for, 128
written statements, preparation of, 130–131
Communication plan, crisis, 101–105
blank media query log sheet, 104
communication training, 102–103
contact lists and key documents, 103–104
create communication flowchart, 102
differentiate communication from notification, 102
establish media staging, 102
identify PIO, 101
maintaining credibility, 104–105
media training sessions, 103
one-page background document for media, 103
research on previous crises, 104
state and disseminate media policy, 101–102
up-to-date media list, 103–104
Community
disaster-resilience, building of, 45–46
in emergency planning, 37–38
led adaptation, 198
led recovery, 198
resiliency
assessment of, 77
reversing trend of declining, 159
Community Emergency Response Teams (CERT) Program, 99, 151
Communitywide recovery, 159
Contingency planning, 43
components of, 39
enables organization to respond, 61
Corrective and preventive action (CAPA) plan, 124
Covello, Vincent, 26
CREAT. *See* Climate Resilience Evaluation and Awareness Tool
Crisis Emergency Risk Communication Manual, 125
Crisis & Emergency Risk Communication Tool Kit for Community Water Systems, 125
Critical Infrastructure Partnership Advisory Council (CIPAC), 23–24
Critical recovery operations, 169–171
Cyberattack, 20
Cybersecurity, 10, 44

D
Damage assessment
Damage Assessment and Reconstruction Monitoring system, 205–207
memoranda of understanding, 170
for recovery, 174
Damage Assessment and Reconstruction Monitoring system (DARMsys™), 205–207
Department of Homeland Security (DHS), 98
best practices, 136
Infrastructure Protection Plan, 138
National Communications System in, 10
National Preparedness Guidelines, 96
National Response Framework, 91–93
DHS. *See* Departments of Homeland Security
Disaster Resilience: A National Imperative, 23
Disaster-resilience community, building of, 45–46
Disasters, 2. *See also* Hazards
definition of, 4
human-made, 18–22. *See also* Human-made disasters
multihazard, 22
natural, 11–18. *See also* Natural disasters
planning/preparedness, 3, 6, 23–25
prevention and mitigation engineering, 54
recovery, 6, 24, 27. *See also* Recovery
reducing risks of, 28–29
response, 6, 24, 25–27
risk elements, 23
Dissemination of information, 77
Drought, 17–18
Dust storms, 17

E
Earthquakes, 12
Ecosystem services, 195
Effective Risk and Crisis Communication during Water Security Emergencies, 129, 130
Emergency Action Plan, from OSHA, 40, 66–70, 90
emergency evacuation procedures, 68
employee alarm system, 69

overview, 66
procedures for operating employees, 68–69
reporting procedures, 66, 68
staff training, 69–70
Emergency conditions, 88
Emergency management agencies (EMAs), 91
Emergency Management Assistance Compact (EMAC), 146
Emergency Management Australia, 46
Emergency management systems, 52
Emergency planning
communities involves in, 37–38
disaster-resilience community, building of, 45–46
elements of, 74–76
Facility Response Plan, 39–41
for family, 38–39
internal and external, 77–78
motivation for, 36–37
NIST Cybersecurity Framework, 44
Oil Removal Contingency Plan, 40, 42
organizations and bodies involves in, 37–39
process of, 78–79
required by regulations, 39, 40
Resource Conservation and Recovery Act (contingency planning), 39, 40, 43
Risk Management Plan, 40, 44–45
SPCC plan, 40, 41–42
Stormwater Pollution Prevention Plan, 40, 42–43
training and community awareness, 45
Emergency response center and operation, 46–54
authority, 52–53
best practices, 47
command and control structure, 48
communication and information management, 49–51
decision-making, 51–52
establish context, 57–58
Incident Command System, 49
integrated emergency management, 47
levels of planning, 47–48
media involvement, 53–54
monitoring and reviewing, 61–62
organization structure, 52
overview, 46–48
risk analysis, 58–59
risk evaluation, 59, 60, 61
risk identification, 58
risk perception, 66
risk treatment, 59, 61, 62, 63–65
vulnerability assessment of community's infrastructure, 62–63
Emergency response plan, development of
community resiliency, assessment of, 77
emergency plan, elements of, 74–76
emergency planning process, 73–74
exercises and training, 78–80
exercising plan, 77–80
hazard effect demands, 73
internal and external emergency management plans, 77–80
overview, 72–73
plan revision, 80
preparedness demands, 72
process of emergency management planning, 78–79
resources and integration of agencies, 76–77
response management demands, 73
simulated or field exercises, 79
standard protocols, 90–91
table-top exercises, 79
telecommunication destruction, 76
Emergency situations, and potential effects, 7–22
human-made disasters, 18–22
infrastructure failure, 7–11
multihazard disasters, 22
natural disasters, 11–18
Employee alarm system, 69
Environmental recovery, 198–199
Extreme temperatures, 17

F

Facility Response Plan (FRP), 39–41
Federal and regional response support
Centers for Disease Control and Prevention, 98–99
crisis communication plan, 101–105
Department of Homeland Security, 98
Federal Bureau of Investigation, 97–98
federal response support, 96–99
Incident Command System, 94–96
incident command team, 101
National Disaster Recovery Framework, 157
National Incident Management System, 94
National Preparedness Guidelines, 96

National Response Framework, 91–93
regional support, 99–101
risk management planning, 100–101
U.S. EPA, 96–97
Water/Wastewater Agency Response Network, 99–100
Federal Bureau of Investigation (FBI), 97–98
best practices, 134
Cyber Division, 137
Federal Emergency Management Agency (FEMA), 15, 174, 186
Emergency Management Guide for Business and Industry, 136
flood maps, 140
and funding of resilient infrastructure, 202
Incident Command System, 6. *See also* Incident Command System
National Incident Management System, 89, 94
phases of emergency response used by, 90
Federal Protective Service program, 98
Federal Water Pollution Control Act (or Clean Water Act), 21
FEMA. *See* Federal Emergency Management Agency
Fire Prevention Plan, from OSHA, 40, 70–72
procedures to control combustible waste materials, 71
regular maintenance of heat-producing equipment, procedures for, 71
significant fire hazards, 70–71
Floods, 14
FRP. *See* Facility Response Plan
Fusion Centers, 98, 139

G
Gayndah Water Supply Intake Station, 208, 209
Geophysical disasters, 12–13
avalanches, 13
earthquakes, 12
landslides, 13
volcanic eruptions, 12–13
Global Facility for Disaster Reduction and Recovery, 199
Government Emergency Telecommunications Service, 10
Guidelines for the Physical Security of Wastewater/Stormwater Utilities, 144

H
Hazardous Response Program, 98, 140
Hazards, 23. *See also* Disasters
and associated incidents/events, 121–122
categories, 120
definition of, 4
natural, 3
Hazards inventory and vulnerability analysis (HIVA), 174
HIVA. *See* Hazards inventory and vulnerability analysis
Holding statement, 126, 128
Homeland Security Exercise and Evaluation Program (HSEEP), 138
Homeland Security Information Network (HSIN), 99, 151
HSEEP. *See* Homeland Security Exercise and Evaluation Program
HSIN. *See* Homeland Security Information Network
Human-made disasters, 18–22
civil unrest and, 19
crime or criminal activities and, 19–20
personnel issues in, 18–19
terrorism and, 20–22
Hurricane Isaac (2012), 15
Hurricane Katrina (2005), 7, 15, 92
Hurricanes, 15
Hydrological disasters, 14–15
floods, 14
mudslides, 14–15
tsunamis and tidal waves, 14

I
IAP. *See* Incident action plan
ICS Form 201, 116
ICS Form 215A, 118
Incident action plan (IAP), 106
preparation and approval, 119–120
Incident Command System (ICS), 6, 45, 47, 49, 94–96, 124
Form 201, 116
Form 215, 118
Form 215A, 118
incident organization chart, 59
organization of, 95–96
principles, 95
roles and positions in incident response, 95
Incident command team, 101
Incident lifecycle, 89

Incident-specific response procedures, 120–124
activation, 123
incident command, 123
incident investigation and corrective action, 124
notification/alarms, 123
on-scene command, 123
other personnel/processes, 123
preparation, 122–123
return to service, 124
Independent Agencies Appropriations Act (P.L. 107-73), 21
Infrastructure failure, 7–11
communications during, 10–11
electrical outages, 9
industrial accidents, 7–9
supply chain during, 11
technological failures, 10–11
transportation accidents, 11
Infrastructure resilience properties, 160
Integrated emergency management (IEM), 47 49
International Federation of Red Cross and Red Crescent Societies (IFRC), 4
International Organization for Standardization (ISO)
22320:2011 standard, 25
31000:2009 standard, 25
Intra-agency communication, 183

K
Key personnel contact list, 103

L
Landslides, 13
Land-use planning, 203
Leadership in Energy and Environmental Design program, 25
Lightning strikes, 15
Localized emergency (low-level emergency), 107

M
Maroochy Water Services incident (2000), 20
Media. *See also* Communication
blank query log sheet, 104
briefing, 131
in emergency plan, 53–54
one-page background document for, 103
respond to questions of, 129
staging, establish, 102
training sessions, 103
up-to-date list of, 103–104
Meteorological disasters, 14–15
hurricanes, 15
lightning strikes, 15
tornadoes, 15
Mitigation
engineering, 54
measures, examples, 63–64
Mudslides, 14
Multihazard disasters, 22

N
NAS. *See* National Academy of Sciences
National Academy of Sciences (NAS)
Disaster Resilience: A National Imperative, 23
issues related to disaster resiliency, 27–28
steps to improve resilience, 25
National Center for Post-Traumatic Stress Syndrome, 141
National Communications System, 10
National Disaster Recovery Framework (NDRF), 157, 158, 183, 185–187
Interagency Operational Plan, 186
Presidential Policy Directive-8, 186
recovery support functions, 186–187
National disaster recovery policy, 185–187
National Emergency Management Association (NEMA), 146
National Environmental Health Association (NEHA), 134
National Fire Protection Association, 98
National Fire Protection Association (NFPA), 148
National Flood Insurance Program (NFIP), 166
National Incident Management System (NIMS), 89, 94, 101
best practices, 136
recovery and, 183
Standards Development, 139
National Insurance Crime Bureau (NICB), 19
National Oceanic and Atmospheric Administration (NOAA), 122
National Pollution Discharge Elimination System (NPDES), 42
National Preparedness Guidelines, 96
National Preparedness System, 186
National Research Council

Committee on Risk Perception and Communication, 55
National Response Framework (NRF), 91–93, 183
National Weather Service (NWS), 16, 122, 123
Natural Disaster Mitigation in Drinking Water and Sewerage Systems Guidelines for Vulnerability Analysis, 149
Natural disasters, 11–18
biological disasters, 11–12
climatological disasters, 17–18
geophysical disasters, 13–14
hydrological disasters, 14–15
meteorological disasters, 15–17
NDRF. *See* National Disaster Recovery Framework
Near-normal operations, 107
NEMA. *See* National Emergency Management Association
NFIP. *See* National Flood Insurance Program
NFPA. *See* National Fire Protection Association
NICB. *See* National Insurance Crime Bureau
NIMS. *See* National Incident Management System
NIST Cybersecurity Framework, 10, 44
NOAA, 15. *See* National Oceanic and Atmospheric Administration
Notification, 102
NPDES. *See* National Pollution Discharge Elimination System
NWS. *See* National Weather Service

O
OASIS. *See* Organization for the Advancement of Structured Information Standards
Occupational Safety and Health Administration (OSHA), 6, 95, 98
1910.38(a) standard, 66
29 CFR1910.165 standard, 69
Emergency Action Plan, 40, 66–70
Fire Prevention Plan, 40, 70–72
personal protective equipment standards, 143
Sandia National Laboratory, 143
Oil Removal Contingency Plan, 40, 42
Operations
period briefing, 120
and warning systems, 74
Ordinance content, recovery, 163
Organization for the Advancement of Structured Information Standards (OASIS), 149
OSHA. *See* Occupational Safety and Health Administration

P
Paper-plan syndrome, 78
Personal protective equipment standards, 143
Photographs, 131–132
PIO. *See* Public information officer
Planning meeting, 119
Planning/preparedness, for disaster, 3, 6, 23–25
PPRR. *See* Prevention, preparedness, response, and recovery
Preimpact preparations, 74
Presidential Policy Directive (PPD)-8, 186
Prevention, preparedness, response, and recovery (PPRR), 192
Protective Security Advisor (PSA), 98, 139
Publication 166, 152
Public information officer (PIO), 101, 102, 183

Q
QFES. *See* Queensland Fire and Emergency Services
QRA. *See* Queensland Reconstruction Authority
Queensland Fire and Emergency Services (QFES), 206
Queensland Reconstruction Authority (QRA), 203, 206–207

R
RCRA. *See* Resource Conservation and Recovery Act
Recovery, 6, 24, 27
community-led, 198
communitywide recovery, 159, 162
coordination and communication, 181–184
communication during recovery, 182–183
communication with public, 183–184
intra-agency communication, 183
National Incident Management System, 183
recovery implementation, 184
environmental, 198–199
implementation, 184

information gathered by RTF and others, 174–176
community needs, identifying, 176
critical infrastructure and key resources
priorities, determining, 175–176
risk assessment, 175
economic effects on community, 176
public health concerns, 176
using hazard analysis information, 175
memoranda of understanding
damage assessment, 170
debris management, 170
State Department of Transportation, 170–171
National Disaster Recovery Framework, 157, 158
operations plan activation, 180–185
damaged infrastructure, inspection of, 180–181
providing support services, 181
recovery coordination and communication, 181–184
ordinance content, 163
overview, 157–159
planning and management, 159–180
administrative duties, 164
community planning process, 164–169
coordination with local plans, 165
critical recovery operations, 169–171
damage assessment, 169–170, 174
documentation requirements and priorities, 169
economic development plans and policies, 166
emergency management plans and policies, 166
finance issues, priorities, and tasks, 168
finance plans and policies, 166
importance of, 162–180
information gathered by RTF, 174–176
legal issues and tasks, 167–168
other local plans and policies, 166–167
public involvement in, 165
public works plans and policies, 165
recovery plan, 171
recovery projects, 176–180. *See also* Recovery projects
recovery task force, establishing, 162–164. *See also* Recovery task force
resource needs, planning for, 171–174
staffing needs, 167
predisaster planning for, 158
priorities
environment, protecting, 179–180
establishing, 174
providing essential services, 178–179
public safety and health, protecting, 179
restoring critical infrastructure and key resources, 179
resources
disasters supported by presidential declarations, 172–173
local recovery funding, 172, 173
reserve funds, 173
risk abatement and risk mitigation, 173
risk acceptance, 174
risk spreading, 173
state and local, 171–172
small-scale disaster recovery, 160–162
training, 184–185
utility integrated contingency plan
advantages of, 161
planning characteristics of, 161–162
primary objectives, 161
Recovery projects
environment, protecting, 179–180
implementation criteria, 178
potential, identifying, 176–177
prioritization of, 177–180
providing essential services, 178–179
public safety and health, protecting, 179
restoring critical infrastructure and key resources, 179
selection to implement, 177–178
Recovery support functions (RSF), 186–187
Recovery task force (RTF)
establishing, 162–164
information gathered by, 174–176
participants, 164
REDI. *See* Repeat Event Dollars Index
Regional emergency
high-level crisis, 110
low-level crisis, 109
Regional response support, 99–101
incident command team, 101
risk management planning, 100–101
Water/Wastewater Agency Response Network, 99–100
Regulatory Emergency Plan Programs, 40
Repeat Event Dollars Index (REDI), 207
Resilience, 27–28
building
in emergency planning, 196
in recovery, 199
in wastewater systems and utilities, 23

building in wastewater systems and utilities, 23
case for, 191–192
definition of, 5, 23, 45–46
in disaster management, 191–199
ecosystem services, 195
interdependencies influences resilience, 194
PPRR model, 192–193
of infrastructure, 199–203. *See also* Resilient infrastructure
integration of
in prevention and preparedness, 196, 197, 204–205
into response and recovery, 196, 198–199, 205–208
to natural hazards, 196
steps to improve, 25
understanding, 192–195
Resilient infrastructure, 199–203
building, 208, 210
communications infrastructure, 200
coordinated approach, 202
design and construction, 203
funding, 202
investment framework, 201
land-use planning, 203
PPRR model, 200
roads, bridges, and other transport infrastructure, 200
strengthening public asset, 201
strengthen resilience, key areas to, 201
water and flood mitigation infrastructure, 202–203
water supply, 200
Resilient Queensland (case stuudy), 203–210
Betterment Fund, 207–208
building resilient infrastructure, 208, 210
damage assessments, 205–207
geospatial overlay of data, 207
in-field, conducting, 206
Gayndah Water Supply Intake Station, 208, 209
Repeat Event Dollars Index, 207
Resource Conservation and Recovery Act (RCRA), 6, 39, 40, 43
Resources and best parctices, 134–143
American Health Organization Regional Office of WHO, 149
American National Standards Institute, 144
American Society of Civil Engineers, 144
Centers for Disease Control and Prevention, 134
Community Emergency Response Teams Program, 151
Department of Homeland Security, 136
DHS Infrastructure Protection Plan, 138
FBI Cyber crime, 137
Federal Bureau of Investigation, 134
Federal Funding for Utilities, 142
FEMA flood maps, 140
Fusion Centers, 139
Hazardous Response Program, 140
Homeland Security Exercise and Evaluation Program, 138
Homeland Security Information Network, 151
Multiagency Response to a Major Water Pipe Break: A Massachusetts Case Study and Evaluation, 152
National Center for Post-Traumatic Stress Syndrome, 141
National Emergency Management Association, 146
National Environmental Services Center, West Virginia University, 135, 146
National Fire Protection Association, 148
National Integrated Drought Information System, 135
National Preparedness Guidelines, 137
NIMS Standards Development, 139
Occupational, Safety, and Health Administration, 143
Organization for the Advancement of Structured Information Standards, 149
Protective Security Advisors, 139
Publication 166, 152
Sandia National Laboratory, 143
U.S. Coast Guard, 135
U.S. Environmental Protection Agency, 141
Vulnerability Self-Assessment Tool, 141
Water and Wastewater Agency Response Network, 145
Water Environment Federation, 150
Water Environment Research Foundation, 150
Water Information Sharing and Analysis Center, 151

Water Infrastructure Security Enhancements, 145
Water Security Publications Database, 141
Resources needs, in recovery planning
disasters supported by presidential declarations, 172–173
local recovery funding, 172, 173
reserve funds, 173
risk abatement and risk mitigation, 173
risk acceptance, 174
risk spreading, 173
state and local, 171–172
Response, 6, 24, 25–27
activities, 26
communications for, 26–27
ISO 22320:2011 standard, 26
ISO 31000:2009 standard, 25
RFC. *See* River Forecast Center
Risk Management Plan (RMP), 40, 44–45, 55
communication and consultation, 55–57
disaster prevention and mitigation engineering, 54
Risk(s), 23
analysis, 58–59
analysis or assessment, 58–59, 175
definition of, 100
evaluation, 59, 60, 61
identification, 58
information checklist, 56
management terms, 100
matrix for scaling consequences of, example of, 65
mitigation measures, example of, 63–64
perception, 66
reduction measures in land use and planning, 64
treatment, 59, 61, 62, 63–65
River Forecast Center (RFC), 122
RMP. *See* Risk Management Plan

S
Safety, in disaster response, 124–125
message/plan, 126, 127
Safety officer, 124
Saffir-Simpson Hurricane Wind Scale, 16
Sandia National Laboratory, 143
SCADA system. *See* Supervisory control and data acquisition system
Skills, used in emergency management, 77
Small-scale disaster recovery, 160–162
Social media, 128
SPCC plan. *See* Spill Prevention, Containment, and Countermeasure plan
Spill Prevention, Containment, and Countermeasure (SPCC) plan, 40, 41–42
Stormwater Pollution Prevention Plan (SWPPP), 40, 42–43
key elements of, 42–43
Superstorm Sandy (October 2012), 7
Supervisory control and data acquisition (SCADA) system, 10–11
Supply chain, during disaster, 11
SWPPP. *See* Stormwater Pollution Prevention Plan

T
Tactics meeting, 118
Terrorism, 20–22
cyberattack, 20
definition by Code of Federal Regulations, 20
Threat
definition of, 100
nature of, 122
Tidal waves, 14
Tornadoes, 15
Training
communication, 102–103
in emergency management, 45
emergency planning, 45
recovery, 184–185
Training, in emergency management, 45
Transportation accidents, 11
Treatment, storage, and disposal facilities (TSDFs), 43
TSDFs. *See* Treatment, storage, and disposal facilities
Tsunamis, 14
Twitter, 126, 128

U
U.S. Coast Guard (USCG), 135
U.S. Environmental Protection Agency (U.S. EPA), 9, 15, 27, 95
approach to response, 97
Climate Resilience Evaluation and Awareness Tool, 97
Effective Risk and Crisis Communication during Water Security Emergencies, 129, 130

federal response support, 96–97
guidance to store chlorine gas, 21
Resource Recovery and Conservation Act permit issued by, 43
Vulnerability Self-Assessment Tool, 97, 141
Water Security Publications Database, 96, 141
U.S. EPA. *See* U.S. Environmental Protection Agency
U.S. Small Business Administration (SBA), 166, 172
Utilities response, 105–124
"during-incident" planning process, 106
incident-specific procedures, 120–124
initial response, 106, 116–120
conducting tactics meeting, preparation of, 118
executing plan and assess progress, 120
incident action plan preparation and approval, 119–120
operations period briefing, 120
planning cycle, start, 116–118
planning meeting, 119
initial response actions, 105–106
planning during incident, 116
response matrix, 107–110
response timeline, 111–115
Utility integrated contingency plan
advantages of, 161
planning characteristics of, 161–162
primary objectives, 161

V

Volcanic eruptions, 12–13
VSAT. *See* Vulnerability Self-Assessment Tool
Vulnerability
definition of, 4–5, 100
infrastructure, reducing, 200–201
Vulnerability Self-Assessment Tool (VSAT), 21–22, 97, 141

W

WARN. *See* Water and Wastewater Agency Response Network; Water/Wastewater Agency Response Network
Wastewater Treatment Works Security Act, 2002, 21
Water and flood mitigation infrastructure, 202–203
Water and Wastewater Agency Response Network (WARN), 99–100, 145
Water Environment Federation, 150
Water Environment Research Foundation, 150
Water Information Sharing and Analysis Center (WaterISAC), 98–99, 151
Water Infrastructure Security Enhancements, 145
WaterISAC. *See* Water Information Sharing and Analysis Center
Water Security Publications Database, 96, 141
Water/Wastewater Agency Response Network (WARN), 99–100, 172, 185
Weapons of mass destruction (WMD), 97–98
Weapons of Mass Destruction Directorate, 98, 134
Wildfires, 17–18
WMD. *See* Weapons of mass destruction
World Bank, 23
connection of three Cs, 28

CPSIA information can be obtained
at www.ICGtesting.com
Printed in the USA
BVHW010801140921
616365BV00016B/5

9 781572 784130